Centennial Farm

How Six Generations Farmed The Land

By Frank Lessiter
and three generations of family members

Lessiter Publications, Brookfield, Wisconsin

Publisher's Cataloging in Publication
(Prepared by Quality Books Inc.)

Lessiter, Frank
    Centennial farm / by Frank Lessiter.
    p. cm.
    Includes index.
    ISBN 0-944079-05-9

    1. Farm life—Michigan—Lake Orion. 2. Family farms— Michigan—Lake Orion. 3. Country life—Michigan—Lake Orion. 4. Lake Orion (Mich.)—History. I. Title.

F572.02L47 1 997   977.4'38'043
                                                    QBI96-40275

International Standard Book Number:
0-944079-05-9

Copyright © 1997 By Lessiter Publications, inc.

All rights reserved.
With the exception of quoting brief passages for the purpose of review,
no part of this book may be reproduced in any form
without the written consent of the publisher.

Published by Lessiter Publications, P.O. Box 624
Brookfield, Wisconsin 53008-0624

For additional copies of this book or information on other Lessiter Publications books or publications, call toll-free (800) 645-8455 or write to the above address.

Manufactured in the United States of America

*Dedicated to
our children, grandchildren, nieces, nephews
and future offspring who wish to learn
how six generations
worked the land in the 1800s and 1900s*

Recollections...

One Family, One Farm,
  140 Plus Years 7

A Wooden Trunk Full Of Dreams 11

The Year Was 1853 17

Family Members Were
  Very Good Record-Keepers 19

The Baldwin Mile 25

Getting Through
  America's Darkest Days 29

Five Generations
  Have Rocked Here 35

When Stagecoaches
  Brought The Mail 37

Lohill—Farm, Drugstore
  And Even A Grocery 39

The Old Tire Swing 43

Farm Mothers Are One Of A Kind ... 47

When Dad Really Got Sick 53

Where's Grandpa? He's Fishing! ... 55

The Family's "Good Child" 59

Friday Nights, Sunday Mornings,
  Chocolate Sodas 65

Oh No! A Snake In The House! 69

Farming Can Be A Big Blast! 71

Trees Really Made The Farm 75

Real Beauty In Farm's Old Barns .. 83

Barn Wood Has Special Meaning 99

Machinery Upstairs,
  Potatoes Downstairs 101

Worrying About The
  "Barn Burners" 103

There's No Place Like Home 107

Beef To Lamb To Milk To Eggs 117

Livestock Tales Of The Past 123

Chicken Tales From Lohill Farm .. 129

Memories of Real Horsepower 131

The Old Farm Lane 137

Haying Wasn't Always
  What It Seemed To Be 139

Those Really Hot
  Days Of Summer 145

Threshers For Dinner—
  In 20 Minutes 147

Conservation Was
  Always Important 155

Mechanical Skills Were Critical 163

Old Tractors Still Going Strong 165

Driving The Old Oakland Cars 169

Hired Hands Essential
  For Farming Success 173

His Name Was Leon McGrath 177

Cold Concrete,
  Late Friday Nights 181

Winter Was Never Easy 183

When Harvesting Actually
  Came In Mid-Winter 185

Pets Add Lots To Farm Life 189

62 Years Without Ever
  Missing A State Fair! 199

Blossom, The Sad
  And Lonely Calf 209

Going With Dad To
  The Detroit Stockyards 215

Bull Rings, Gilt Rings,
  Threshing Rings
  And Other Rings 221

When Sundays Were Very Special ..223

Putting On Your Sunday Best 229

Christmas Farm Memories 233

The Old "What-Not" Shelf 237

When Hollywood Came Calling 239

The Murder At Lohill Farm? 243

1957 And 1958...B.D. And A.D. 245

Corn Was King 249

Lake Orion's Boys Of Summer 251

Our Role In Rural
  Newspaper Delivery 255

SIX GENERATIONS OF MEMORIES. Stories about the family and farm were handed down from generation to generation. The source of many of the recollections found in this book was John Lessiter, the fourth generation of the family to live on the farm.

Hi-Yo Silver—Awaaaayyyy! 263

Farm Families
  Relish Good Friends 267

Texas-Style Farm Vacations 273

It Doesn't Get Any
  Better Than This 275

Special Letters To
  Grandma Lessiter 285

Township Service Was
  Long-Time Family Tradition 293

Family's Second
  Centennial Farm 297

Mailboxes At 501 300

The Canadian Farm Family 301

Visiting The Old
  Taylor Farm In Canada 307

The Farm Next Door 309

One Of America's
  Best Dairy Herds 319

When Ford Didn't
  Have The Bucks 327

When Dad Served
  On The School Board 331

Everything I Ever Learned 333

82 Reasons Why Farming
  Is The Best Way Of Life 343

45 Things I Remember Most About
  Growing Up On Lohill Farm 347

Family Favorites 351

Orion Township—1819 To 1880 355

200 Years Of Family History 357

The Lessiter Family Players 365

5

Lohill Farm...

Taken in the early 1950s, this aerial photo of the Lessiter Centennial Farm shows the family farmstead located along Baldwin Road, five miles west of Lake Orion, Michigan.

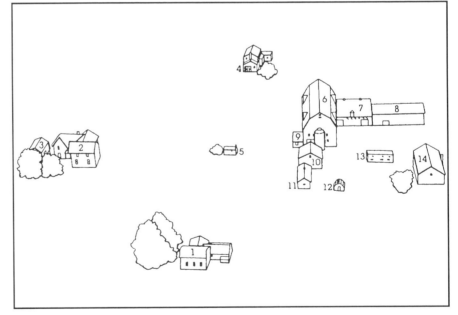

1. Original farm home and tenant house.
2. Norah and Frank Lessiter home.
3. Wood shed, ice house and garage.
4. Donalda and John Lessiter house.
5. Old milkhouse.
6. Cattle barn
7. Heifer barn.
8. Sheep barn.
9. New milkhouse.
10. Corn crib and machinery storage.
11. Farm shop.
12. Brooder house.
13. Hen house.
14. Truck garage and potato storage.

One Family, One Farm, 140 Plus Years

SIX GENERATIONS OF our family have lived on the same farm near Lake Orion, Michigan, since 1853.

About two dozen years ago, family members started jotting down the many pleasant memories we had of life on the farm. Our goal was to share these experiences with our children, grandchildren, friends and many others so that they knew how our family has lived and operated this farm since its beginnings way back in 1853.

Over the years, we've shared many of the early day experiences Dad and Mom remembered and also the stories that Dad had been told by his parents and grandparents.

Many of these stories are probably similar to those you've been told by your parents and grandparents or have shared

MANY MEMORIES. Janet and Frank join parents, Donalda and John Lessiter, for fortieth wedding anniversary in 1979. Many memories of Lohill Farm in this book came from these four and other generations of this Centennial Farm family.

A TRIBUTE TO LOHILL FARM...

WE DRIVE UP the gravel driveway early in the morning.

The sun is slowly peering out from behind the green rolling hills. The long blades of green grass still drip with morning dew and we are greeted with the loud "good morning" crow of the rooster.

Back Home In Michigan

We are at Lohill Farm in southeastern Michigan, where my father grew up. Lohill is a Centennial Farm, since it has been owned by the Lessiter family for over 140 years. As a result, Lohill Farm holds many special memories for me.

We drive up the winding driveway to the simple, white farmhouse. We enter the screened in porch and see the immaculate green-carpeted parlor with comfortable chairs, sofa, a few antiques and a giant wooden piano.

Walking up the tall, creaky stairs, we see the room where my Dad slept, waking up from spring until fall to sunshine bouncing off the walls before he headed to the barns to do chores.

Downstairs in the kitchen where we can smell bacon frying, we are greeted with a smile from Grandma, rocking and reading contently in her chair.

After a hearty breakfast of bacon and scrambled eggs, we leave to tour the rest of the beautiful farm.

The first thing you notice is the big white, wooden barn— the kind you often see in paintings. Inside, hay and straw bales are stacked high, as particles of dust float down through the shadows. We climb into the dark loft and see the yellow patch of straw below. We can hear the faint sounds of the cows and chickens.

140 Years Of Hard Work

As we walk from the barn, we see the long, gravel lane leading back to the edge of the property. The lane has been embedded into the green pasture, with over 140 years of steady, difficult work.

At the end of the lane is a small, round and deep lake—dropping off to a 60-foot depth just 3-feet off shore. The kind of place most farmers would like to call their own, this lake creates a sense of mystery to the land and is serene and calm. No one else is around.

As we walk back to the barn, we see a towering oak tree holding a single tire swing by a worn, tired rope. You sit in the tire and let the cool breeze blow through your hair.

Lohill Farm...it will always be a very special place to me.

—*Mike Lessiter*

---

with your own children and grandchildren.

This book is about the many happenings at our own Centennial Farm (actually, there are two Centennial Farms in the family, but more on that later in this book). It includes many memories which took place at Lohill Farm over the past 140 plus years.

As family members have grown older, the most vivid

THREE GENERATIONS. Mike, Frank and John Lessiter take a break on a farm building project.

BIG CHANGES. In this aerial photo taken in the mid 1990s, the Lessiter farmstead appears in the upper righthand corner. As late as 1965, only five houses would have appeared in an aerial photo of the Dennis Lake area. In this photo, 106 homes are shown, indicating the drastic change which took place in this area from farming to suburbia.

SOIL MAPS. The farm's Soil Conservation Service maps show Osthemo Boyer loam sand on 6 to 40 percent slopes, Houghton and Adrian muck soils, Riddles sandy loam on 12 to 18 percent slopes. Others soils include Fox Riddle sandy loam on 1 to 12 percent slopes, Fox sandy loam on 6 to 12 percent slopes, Sebewa loam, Gilford sandy loam, Spinks on 0 to 8 percent slopes and Omas loamy sand on 0 to 12 percent slopes.

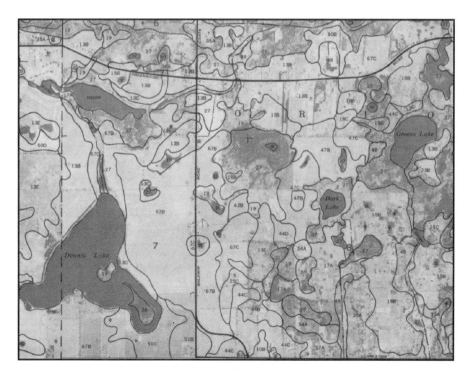

CENTENNIAL FARM, 1853-1953

Today the old farm we proudly acclaim,
For a Centennial Farm it became.
In the annals of our illustrious state.
Its long history was recorded on this date,
A history full of legends of yore
Of hardships, heartbreaks and joys galore.

Many changes have come as changes do.
And there is much to be said for the new.
The cows no longer are milked by hand,
And the horses no longer plow the land.
Machines do all the chores and work today,
And large-scale production has come to stay.

Mother hen no longer sits on her nest.
Eggs must be graded and come up to test.
Poultry is fed to make meat dark or light,
And hogs are raised to please the appetite.
Milk must come up to regulation.
All are part of the farm renovation.

But for all these changes and modern trends,
The farmer still stoops and his back still bends
From his countless and never-ending chores,
As he scorns and good-naturedly ignores
The modern 8-hour day and 5-day week
Which the city worker chooses to keep.

The farmer has freedom and peace of mind
So precious and sought after by mankind.
A sense of accomplishment is his reward
As he achieves the goals he's worked towards.
He's happy with his chosen way of life,
Contently free from city noise and strife.

As I look out over the fields of green
And see the beauty of our country scene,
I think of those who settled here in the past
On this land where they chose their lots to cast.
How they dreamed, planned and sacrificed, too
And worked hard to make visions come true.

This farm has seen many men come and go.
It has seen progress—some fast, some slow.
It has seen men put their roots in this soil,
Nurtured with the sweat of their honest toil.
Godly and upright these men of the sod
True to themselves, but above all, to God.

Today we point to the old farm with pride
To the legacy left by those who've died,
To a heritage rich beyond compare
Entrusted to us to cherish and to share.
May we prove worthy to the task each day,
And may God bless us all, in His dear way.

—*Donalda Lessiter*

---

memories of life on the farm have centered about pleasant events and happy days.

The memories in this book preserve much of which our family has learned to cherish, the important items that we have built our lives on, the local history that we have lived through and the heritage we wish to pass on to future generations of the family.

These favorable memories, experiences and recollections of Lohill Farm are of special interest to our children and grandchildren.

But they also have special meaning for anyone who has an interest in the ever changing American farming lifestyle as these stories are similar to those which many other families have experienced over the years.

We hope you enjoy these stories from one of America's Centennial Farms as much as our family members have enjoyed putting them together.

—*The Lessiter family*

A Wooden Trunk Full Of Dreams

YESTERDAY MORNING I was cleaning my cluttered office and took a long look at the old cedar trunk sitting on the floor in front of my window.

I use this trunk, its interior covered with British newsprint from 1847, as a "filing cabinet" to hold decades of important farming articles.

This trunk is the same one in which my Great Grandfather carried all of his belongings when he made the 6-week voyage to America to follow his dreams. With his trunk in my office nearly 150 years later, I realized I'm following mine.

In 1847 in Liverpool, England, a young man kissed his sweetheart good-bye, threw a trunk on board and left on a ship sailing to America with his youngest brother.

He was my Great-Grandfather, John Lessiter, a native of Wiltshire, England, who came to America when he was 20 along with his 14-year-old brother,

THE FAMILY TRUNK. All of John Lessiter's belongings were in this trunk when he left Liverpool, England, in 1847 for the 6-week journey to America and the eventual move to Michigan. Shown here is his namesake and grandson, John Lessiter.

FULL OF FAMILY HISTORY. For over 100 years, the old family trunk was filled with valuable family heirlooms such as the family Bible brought from England, old photos, old newspaper clippings and much more. It included a "goldmine" of Lessiter family history.

James. His father, William, had settled near the village of Orion, Mich., 4 years earlier, and the two sons set out to join him.

Besides James, John had two younger brothers and an older sister Elizabeth, who died when she was 8. Two other sons, Henry and William, also later settled in Michigan.

John's mother, Elizabeth Kington, died when he was five, and his father married Elizabeth Sheppard shortly afterward. John grew up on an English farm and went to public school until he was 12. Later, his father kept him in boarding school.

When he left for America at 20 years of age, John promised his sweetheart, a dressmaker by trade, that he'd go back for her. He never saw her again.

When he arrived in America, he located his father and worked month by month on an Orion farm. He later bought his employer's livestock and rented the farmland for a number of years. He soon met a Clarkston, Michigan, woman named Nancy Beardsley and was married on New Year's Day, 1853.

John and Nancy saved enough money to buy 120 acres in Orion township in 1853. Only 30 acres of this land had been broken—the remainder was still wild. He built a small frame house and began clearing the land, then started raising and breeding Shorthorn cattle in 1858.

In 1865, John brought five Durham cattle from New York, introducing the first purebred stock in the township. At its peak, the farm totaled 436 acres.

When he died in 1901, a local obituary, titled "Another Pioneer

> *"He started with nothing and at the time of his death was one of the wealthiest pioneer farmers..."*

Gone," reported, "...he started with nothing and at the time of his death was one of the wealthiest pioneer farmers of the county and one of the most truly respected and influential citizens."

Starting A Family

My Grandpa Frank was the fourth of John and Nancy's six children, born in 1862.

In his early years, he went to the nearby Block School. He never went to high school, but completed a course in bookkeeping at the Business College of Pontiac when he was 21. He started working on his dad's farm at the bottom—for just $150 annually for several years.

After his father's death in 1901, Grandpa Frank and his brother, Floyd, bought the farm. Grandpa Frank became one of the leading stockmen of the township, raising registered Shorthorn cattle.

A 1912 historical volume referred to his land as, "a beautiful farm, with large maple and pine trees bordering the road and a large lake at the rear of the house that further adds to the home's comfort.

"In addition to the large, modern steam-heated house, there is also a tenant house, which was the first home built by his father and which has been remodeled."

Grandma Norah and Grandpa

FOUR GENERATIONS. In this 1909 photo, left to right are: Lucy Everett, Norah Lessiter, John Lessiter and Mary Elizabeth Everett Wiser. Grandma Wiser (shown at far right in the above photo) is shown here at her 100th birthday party in 1945.

Frank were married in 1895. A daughter, Caroline, died in infancy and their only surviving child, my Dad, Milon John, was born in 1908. Dad married Donalda Taylor in 1938. I was born in 1939 and my sister, Janet, was born in 1946.

Orion's Early Years

Incidentally, Grandpa Frank's mother was the granddaughter of Moses Munson, Orion's first settler, who built a log cabin in the area around 1825. Munson also built the first sawmill later that year, boosting further settlement.

Early settlers followed, mostly from New Jersey and New York. After clearing land, many settlers planted orchards.

Reaching Orion to start a new

NORAH AND FRANK. They were married in 1895 in the Seymour Lake Methodist Church which was located near the Wiser family Centennial Farm.

GRANDPA FRANK. A leader in the community, he was one of the descendants of the Lessiter family who came from England and started the home farm in Lake Orion in 1853.

GRANDMA NORAH. During her farming lifetime, she lived and worked on two Centennial family farms located only 7 miles apart.

life was no easy task in those days. The only road was a trail through the woods barely big enough for a wagon and oxen to pass.

Many came to Detroit by steamer, bringing a few household items to make a new home in the wilderness, which led them over impassable roads through forests. Some even made the long journey on foot, wading through knee-deep mud and trying to find their way through thick forests which were common in the area.

In 1837, a neighboring town known as Deckerville moved its stores and offices to what is now known as the village of Orion. A post office was later established at the farm home of John Lessiter, located in a section known as Jersey, and it continued there for 30 years.

As postmaster, his entire salary was paid from the sale of stamps. Mail was first carried by horseback and later by stage coach from Pontiac to Lapeer. John also served as justice of the peace for 18 years.

Rest Of The Family

Great Grandma Wiser (Mary Elizabeth Everett): She and her husband, Milon, owned the other Centennial Farm in the family at Seymour Lake which is about 7 miles from the home farm. She lived as a widow for 53 years.

Grandma Norah used to make her eggnogs and spike them with whiskey. Great Grandma disapproved of liquor, but never knew she was downing it daily.

Grandma never measured out the whiskey—sometimes it was pretty stiff. "We always said your great grandma would turn over in her grave if she knew how much whiskey was in those eggnogs," laughs Dad.

Shortly after my parents were

GRANDMA TARP. A Michigan native, she lived in Alberta before moving to Detroit and eventually to Lake Orion, Michigan. She's shown here with her first grandson, Frank, on the front steps of her daughter Donalda's new home on the Lessiter Centennial Farm.

3 months after the family celebrated her 100th birthday.

Grandma Norah (Wiser) Lessiter: She lived across the road, and if Mom wouldn't let me have any candy, I'd go over to Grandma's and eat all I wanted. She loved cards and we would play rummy for hours.

One thing that always miffed my parents about Grandma Norah was the annual Eastern Star Christmas Party.

At this annual party, parents brought one gift for each of their kids and put it under the tree. "Santa Claus" read each child's name and he or she would run up front and excitedly tear through the wrapping paper.

married, Grandma Wiser told Mom to be sure and take good care of my Dad, because many die young in the Lessiter family. She was in her 90s then.

Mom always thought that was rather funny. Ironically, Great Grandma lived until late in 1945,

GRANDPA TARP. A great fan of major league baseball and crossword puzzles, Grandpa Tarp is shown here at a family Christmas celebration with his daughter, Donalda, and Dietrich Ristow, a German exchange student who spent a year living with the John Lessiter family in the mid-1950s.

TWO GENERATIONS. John Lessiter and his father, Frank Lessiter, are shown in the early 1930s standing on the front lawn of the big family house. Dressed up for a family picnic, both men are wearing ties—one of the long term traditions of the Lessiter family when it came to heading for town or entertaining bull buyers, relatives or friends.

Year after year, Grandma would sneak a gift for me under the tree and I was always the only kid in town to get two gifts. Every year, my folks pleaded with her to stop and she'd agree. But when the names were read, mine was always called twice.

Grandpa Frank Lessiter: Besides farming, he loved to fish and shoot pool. When he and Grandma went to town, he'd stop by the pool hall and play a few games.

His bank lent my parents money to build their house and he was always concerned about their expenses. Once he was harping to my dad about the more expensive colored bathroom fixtures my mother wanted. Mom finally told Dad, "Who's going to live in this house, your Dad or me?" She got the colored bathroom fixtures.

Grandma (Janet Taylor) Tarpening: Her first husband, Donald Taylor, died of a heart attack 15 days before my Mother was born. After farming for several years, he had operated a livery stable in Nanton, Alberta.

She named my mother Donalda, after the father she never had the chance to meet. Grandma Janet moved back to Detroit, Michigan, and married Sherman Tarpening in 1919.

When my parents played cards on Saturday nights, they usually dropped me at Grandma Tarp's house to spend the night.

We always made popcorn on the old gas stove. From 8:00 to

"Farm families are often closer than any other kind of families ..."

8:30 we listened to "Truth or Consequences" on the radio. She also worked as a matron in the county jail for female prisoners.

Grandpa Sherm Tarpening: Grandpa Tarp was really an "adopted" grandfather to me, and a super guy. No one ever would've guessed he was my Mom's stepfather. He loved listening to Detroit Tigers baseball games and doing crossword puzzles—he could do most of them in minutes.

When he worked the night shift at the greenhouse, sometimes I'd go with him while he tended the boilers, flowers and plants.

He drove a great old car (1941 Ford—I don't remember the model) which, incidentally, led to an ugly incident in his garage. His cat liked to sleep under the hood in the winter, and one day when Grandpa started the car up, the cat got caught in the fan.

Tell The Stories

Farm families are often closer than any other kind of families in the world. There's a strong sense of values and families tend to stay close to home and make a living together.

This year, after one of the big family get-togethers, such as Thanksgiving or Christmas dinner, sit down with your children and grandchildren and tell them a story or two about family members who've passed on. Let them know who they were and how their ancestors struggled to make a better life for them today.

Years ago when I was in England, I looked through the phone book and called some of the Lessiters there. I found I must be related to some candy producers, since I ran into Lessiter Chocolatiers candy stands, which were located throughout London and some British towns.

When I told them about William, John and some of the others that made their way to Michigan, they said they didn't recognize those names.

Those days were a long time ago, but maybe they'd heard about me and how I would sneak over to Grandma Norah's for candy. Maybe they just thought I was a sweet-toothed Yankee looking for a couple chocolate nonpareils.

The Year Was 1853

THERE WAS PLENTY of excitement in the world in 1853, the year when the first members of the Lessiter family settled at the farm in Michigan.

To give you a better understanding of what pioneer life meant in those days in the Lake Orion, Michigan, area, here's a rundown on a few of the world news highlights from 1853:

Australia. Melbourne University is founded.

Austria. First railroad is built through the Alps, going from Vienna to Trieste.

Belgium. The first International Statistical Congress is held in Brussels.

Connecticut. Protesting a requirement that women cover their legs, Amelia Jenks Bloomer, an advocate of women's rights, gives a speech wearing a pair of Turkish-style pantaloons under a short skirt.

France. Napoleon III marries Eugenie de Montijo.

Germany. Austria and Prussia sign 12-year business treaty.

Great Britain. Vaccination against smallpox is made compulsory.

Great Britain. Peace is achieved between Britain and Burma.

India. The first railway opens linking Bombay and Thana.

Indiana. The 459-mile long Wabash Canal is completed.

Japan. A U.S. squadron led by Commodore Matthew Perry arrives in Tokyo harbor and de-

Farm Prices In The 1850s...

Product	Price
Corn	66 cents per bushel.
Wheat	$2.06 per bushel.
Oats	47 cents per bushel.
Barley	95 cents per bushel.
Rye	$1.05 per bushel.
Buckwheat	94 cents per bushel.
Irish potatoes	$1.11 per hundredweight.
Sweet potatoes	$1.93 per hundredweight.
Hay	$14.48 per ton.
Wool	22 cents per pound.
Milk	$1.58 per hundredweight.
Chicken	18 cents per pound.
Eggs	23 cents per dozen.
Turkeys	24 cents per pound.

—*United States Department of Agriculture*

mands that Japan open itself up for trade with the outside world.

New Orleans. Yellow fever epidemic kills 11,000 people.

New York City. The show, "Uncle Tom's Cabin," opens.

New York City. Henry Steinway opens piano factory

Niagara Falls. John Roebling finishes building the Niagara suspension bridge from New York to Canada.

Oregon Territory. Table Rock Treaty is concluded in which Indians agree to transfer land to settlers for $600,000. *(But the payment is never made).*

Oregon Territory. Williamette University is chartered, the first university west of Rockies.

Sacramento. Selected as California's state capital.

San Francisco. The country's first Buddhist temple, Kong Chow Temple, is built.

Scotland. Rebuilding of the ancient Balmoral Castle at Aberdeenshire begins under the direction of P.C. Albert.

Washington, D.C. Franklin Pierce is inaugurated as the 14th President of the U.S., reciting inaugural speech from memory.

Washington, D.C. Coinage Act of 1853 passes, cutting the quantity of silver in coins under $1 dollar and authorizing $3 gold pieces.

Wisconsin. Norwegian immigrants found the Norwegian Evangelical Church of America.

Early Farmers Watched Costs, Profits

The cost of producing a crop and the resulting net income were just as important to farmers 150 years ago as they are today.

Balance sheets were common for many farmers in the days around when the Lessiter farm was started in 1853.

In the early 1850s, a Cheshire County, New Hampshire, farmer listed his costs of growing corn at $36 per acre:
- $2 for plowing.
- 75 cents for harrowing.
- 25 cents for furrowing.
- $20 for 20 loads of manure.
- $2 extra for putting manure on the hills.
- $1 for planting corn.
- $4 for hoeing the crop twice.
- $1 for cutting the corn.
- $4 for husking.
- $1 per acre for harvesting.

A strong believer in conservation benefits to his land, the farmer valued the corn fodder at $10 per acre and the "paper value" of the remaining manure in the soil at $10, bringing his net cost to $16 an acre.

His excellent yield was 75 bushels per acre. At 70 cents per bushel, this represented a gross income of $52.50 per acre.

Deducting his "calculated paper costs" of $16 an acre left the farmer with a net income of $36.50 an acre. Or a true net income of $16.50 per acre when all out-of-pocket costs were considered.

WINTER WHEAT (Rickenbrode).

This is a new variety of winter wheat which was first discovered by J.F. Rickenbrode of Westfield, N.Y., in 1875. Subsequently grown by him, the variety became known by his name.

It is a white smooth wheat with stiff straw, large heads, well-filled heads, not liable to rust or winter kill; it is prolific and is said to average 30 bushels per acre. It is also excellent for milling and makes a white flour of superior quality.

You will carefully note the time of sowing, whether in drills or broadcast, kind of soils, habits of growth, time of ripening, its yield and report results promptly to this department. This is required of all persons receiving seeds as a condition of the distribution.

—United States Department of Agriculture Crop trial notes in old Lessiter family records.

Family Members Were Very Good Record-Keepers!

IT MAY BE battered and bruised, the pages are yellowed, a few sheets are torn loose, the cover has seen better days and the ink and penciled notes made by early-day members of the Lessiter family are fading.

But the remarkable thing is this old farm accounts book actually exists at all.

Some 135-plus years after the first penciled entry was made in 1861, this now highly-fragile farm accounts book which outlines the farm's early-day income and expenses still survives.

Through a number of genera-

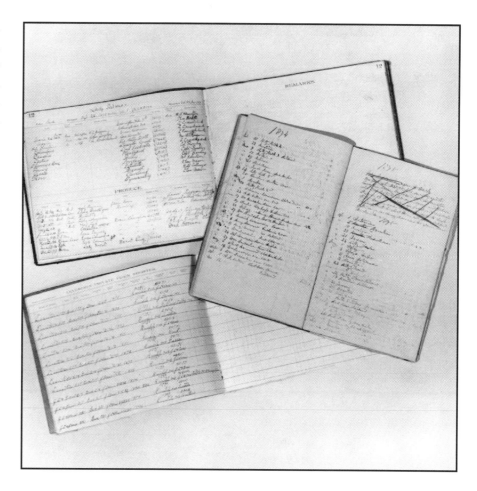

22 YEARS OF HISTORY. Detailed records were kept in this farm accounts ledger from 1861 through 1883 by family members.

EARLY FARM HISTORY. Records entered by the first of six generations of the Lessiter family are examined by Molly Hansen.

tions, the old records book has somehow been preserved without being housed under glass in a museum's surroundings.

Actually part of this book's real charm comes from the hundreds of pencil and ink items written in several different writing styles in the mid 1800s. They deplict much of the farm's unique history.

Over the past 135 years, the well-worn book covering the economical side of farm life from 1861 until 1883, has been carefully studied by a number of folks who found it filled with plenty of Lessiter farm history.

1861...

March 11: William Davis commenced work at $12.00 per month.

March 20: Medicine from Doctor Earl was 31 cents.

March 29: Pair of shoes purchased for $1.50.

1862...

March 1: Elijah *(no last name was given)* commenced work and agrees to work for 8 months at $12 per month.

November 17: Thomas Kay commenced work and agrees to work for 1 year for $132.00.

1863...

April 25: Purchased 5 quarts of clover seed for $1.00.

April 30: 1 bushel of oats sold for 58 cents.

May 2: 2 bushels of wheat sold for $2.60.

June 20: Bought 1 pair of boots for $4.80.

September 1: Mary Close commenced work and agrees to work for 1 year at 10 shillings per week.

1864...

January 12: Paid 20 cents for subscription to *Country Gentlemen*.
March 2: Bought 30 sheep for $160.00.
April 1: Sold seven head of cattle for $460.00.
June 21: Paid 6 cents for 1-year subscription to *Michigan Farmer*.
July 1: Sold one heifer for $40.00 and 42 lambs for $110.00.
September 12: Paid $19.00 for the threshing of 233 bushels of wheat, 123 bushels of barley and 79 bushels of oats.

1865...

March 1: Purchased pair of oxen for $250.00.
March 26: Sold 34 sheep for $277.50 and nine head of cattle for $405.00.
April 1: Sold one horse for $175.00.
April 1: John Lessiter sold one cow, Kitty Kirk, for $13.25.
May 5: Sold potatoes for $14.60, wool for $100.00 and a calf for $40.00.
July 2: John Lessiter sold one cow, Red Native, for $13.75.
July 27: Bought 1,000 pounds of bran for $7.00.
July 29: Bought soap, brush and buttons for 30 cents.
September 16: Sold 26 bushels of wheat ($1.90 per bushel) for $49.40.
September 19: Sold one pig for $2.50.
October 12: Sold sheep for $20.00, hogs for $26.00, wheat

PRIVATE FLOCK REGISTER. Purchased for $1.25, this record book from Mortimer Levevering in Lafayette, Ind., was used to record data on 120 sheep born on the farm between 1888 and 1897. The book's introduction says it is "For keeping records of imported and American bred sheep in a short, concise method." Checked out here by Ryan Hansen, the book had room for lamb data in 20 categories.

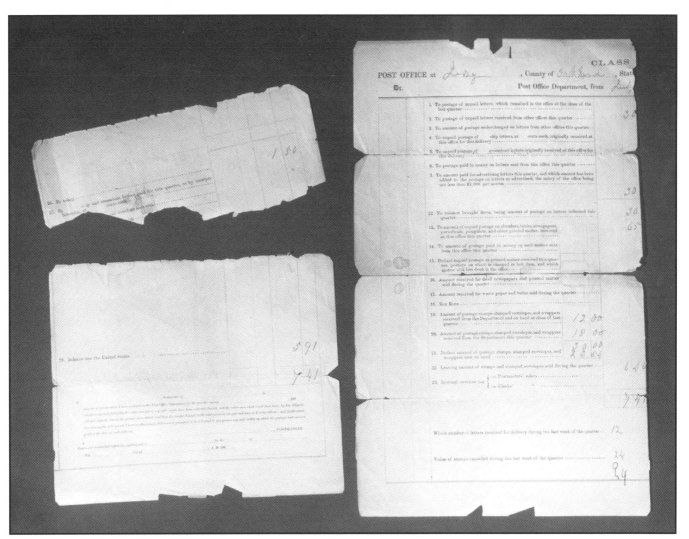

OFFICIAL POSTAL RECORD. The area west of Lake Orion was served by the Jersey Post Office. It was located in a room of the Lessiter family home from 1854 to 1887. These well-worn records show the business recorded in the post office during the third quarter of 1869. During July, August and September, $6.46 of stamps were sold to area residents. The U.S. government was owed $7.41 to cover the various mail charges.

for $52.00 and oats for $56.02.

October 28: Paid threshing crew $16.31.

December 15: 6 days of road work in 1865—$13.75.

1866...

March 3: Sold a bull calf to Jake Burton for $40.00.

March 8: Sold 19 sheep for $116.00.

April 3: Bought 2 tons of barn plaster for $19.50.

May 3: Purchased plough castings for $6.50.

May 17: Sold 4 bushels of beans at 65 cents per bushel or $2.60.

1867...

April 1: Received of John Lessiter $1.00 in full payment for labor performed in Orion township.

1868...

January 3: Paid Nelson Wells $175.00 for five steers.

July 18: Sold one heifer to John Gingel for $27.00.

October 10: Paid George Creamer $19.00 for fat hog.

October 29: Paid Joseph England $35.00 in full for 5 lambs.

November 5: The cow, Kitty Kirk, was returned from the buyer *(this was 3 years after the purchase)*.

November 21: Paid $1.25 for watch.

November 21: Paid 25 cents for camomile and 10 cents for paper.

1869...

January 11: Paid 40 cents

for tobacco.

February 2: Sold 30 bushels of wheat for $45.00.

February 22: Sold one calf for $100.00.

February 26: Bought 3 bushels of clover seed for $27.00.

June 27: Bought 200 pounds of feed for $3.00.

July 6: Bought 9 cents worth of stamps for letters.

August 3: Sold 30 bushels of corn for $22.50.

1870...

May 23: Sold 2 bushels of wheat for $2.80.

June 15: Bought cough medicine for 25 cents.

June 27: Purchased pair of shoes for $2.50.

July 27: Harry Bond worked at harvesting for 10 days and was paid $20.00.

1871...

May 9: Paid 75 cents for a year's *Detroit Free Press* subscription.

June 7: Paid Charles Stones 50 cents for 1/2 day of spreading manure.

LOTS OF HISTORY. Even though these official herdbooks of the American Shorthorn Association cover the years from 1907 through 1919 and help tell the story of the early-day beef program at the Lessiter family farm, they didn't go that far back in the history of this breed.

In fact, one of the 1907 herd books in this collection was actually volume 70 of the official association herdbooks.

June 12: Bought lime for house at 50 cents.

1872...

January 1: Samuel Gomlinson commenced work today for 1 year at $175.00.

June 7: Sold rooster for 35 cents, chalk for 25 cents and pigeons for 25 cents.
June 28: Bought 1 bushel of peas for 75 cents.
December 24: Paid 15 cents for mending boots.

1873...

March 1: Paid house rent for 1 year of $40.00.
July 12: Worked 3 days in hay for $3.75.

1874...

February 15: Paid 60 cents for whiskey.
March 21: Abraham Conant agrees to work commencing about the first of April at $22 per month, to have rent-free house and pasture for 1 cow for the term of 7 months. Conant is to pay for his board and what meals he eats here at $4 per week for 21 meals.

1875...

March 15: Sold two sheep for $27.00.
December 17: Bought pair of mittens for $1.00 and 50 cents worth of tobacco.

1877...

July 1: Purchased one pair of shoes, two pair of socks and one straw hat for $3.05.

1879...

April 3: Bought pair of overalls for 75 cents.

1882...

November 9: Bought 26 pounds of pork for $2.08.

1883...

May 6: Bought 100 pounds of flour for $3.00.
December 7: Purchased 2 gallons of oil for 20 cents.

SHORTHORN HISTORY. A number of the farm's outstanding sires show up in the official herd books of the American Shorthorn Association. They make for most interesting reading in retracing the history of Shorthorn cattle on the farm.

The Baldwin Mile

PLENTY OF CHANGES were made in southeastern Michigan farming operations from 1853 to the mid-1990s. This was certainly the case along Baldwin Road.

Based on old records and recollections of family members, the next three pages indicate the changes that occurred over the last 100 years on a one-mile stretch of Baldwin Road.

In the 1895, 1994 and 1995 drawings of this one-mile-long stretch, Baldwin Road runs through the center of the page from top to bottom. The road at the top of the page is Indianwood Road, while the road across the bottom is Clarkston Road.

The original house on this stretch of road was the Lessiter family home built in 1854. Still standing today, the house (shown with three windows located on the south side of the structure) appears near the center of the page along the east side of Baldwin Road. In later years, the Lessiter farmstead was built north of this house.

In 1895, only six homes and a dozen barns and outbuildings were located along this stretch of Baldwin Road. While some areas were being cropped, there were still a considerable amount of wooded areas. The southeast portion of Lowry Lake is shown in the lower left-hand corner.

By 1945, the area was becoming urbanized. There were a dozen homes and 20 barns and outbuildings. Corn, oats, wheat, barley, alfalfa and other forages were grown on the land. For many years, the lake had been known as Dennis Lake.

By 1995, the land was no longer being cropped and homes were sprouting everywhere. The lake is now called Heather Lake and became the central focus of a stylish subdivision.

BIG-TIME CHANGES. In the old days, the Baldwin mile was a seldom-traveled wheel track-rutted lane. Today, it's a very busy stretch of asphalt and a major north and south linkup in Oakland County.

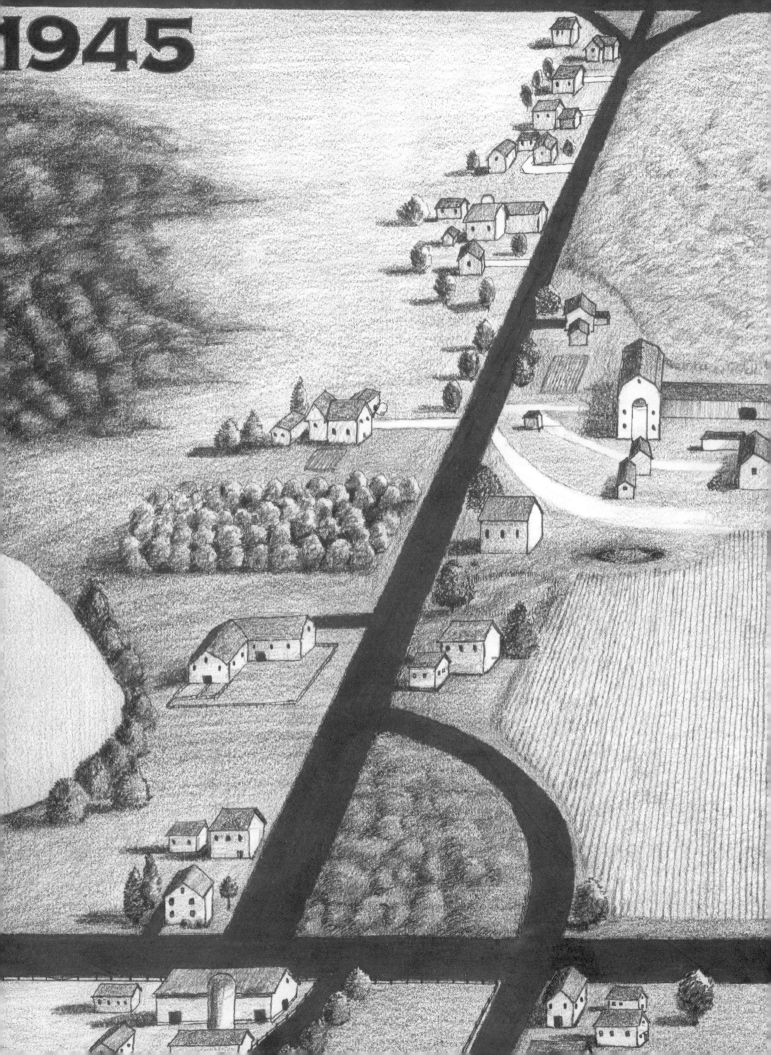

1995

Getting Through America's Darkest Days

FARMERS ARE basically the biggest gamblers found anywhere in the world. There's really no other way of life in which success and survival are virtually out of your hands and in those of Mother Nature, the economy, the bank and so many other things.

Either things work out, or they don't. It's that simple.

Year after year, farmers face an array of serious problems—struggles with banks, droughts, freezes, low market prices and many others.

When times seem unbearable, others might throw in the towel and change jobs, but not farmers. They just dig their feet in and work even harder.

Virtually every farm located in the U.S. has fallen on hard times at some point in time, and

1918 SUGAR RATIONING CARD. During World War I, this sugar ration card was used by Grandma Norah to obtain 12 pounds of sugar in September, 12 pounds in October and 18 pounds in November for a half dozen people living in the big farmhouse. The punched holes indicate sugar was purchased 14 times during the 90-day period. Approved for use at Chase's grocery department, this card had to be presented each time sugar was purchased.

Sugar for canning purposes was purchased in accordance with the government's canning regulations and not with this card. The card was the property of the Lessiter family and sugar had to be obtained at the same place each month during the World War I rationing period.

29

CHANGING NEEDS. Built early in the 20th century, the old barn at right was originally used as a hog and sheep barn. It was later used for hay storage and feeding the loose-housed dairy herd. The white building at left was built in the late 1940s and was used to house heifers and bulls. In the late 1950s, it was converted into a 1,000-bird laying house. In the foregound is the old henhouse which held about 75 birds.

our family's Lohill Farm was certainly no exception to this rule.

Have Degree, Need Work

Dad graduated from Michigan Agricultural College (later known as Michigan State College and later as Michigan State University) with an ag engineering degree in 1931—right in the heart of the Great Depression.

Upon returning home, he couldn't find a job and went to work on the farm, which at the time was raising beef cattle.

"There just weren't any jobs anywhere," he recalls.

As a young man in the 1930s, Dad remembers the Great Depression vividly. But even before that, the severe farm depression of the 1920s had taken its toll.

Farm market prices were so low for beef in those days that the farm switched to a dairy operation, which seemed to be the trend for Detroit-area farmers.

In the years following the 1929 stock market crash, however, there was no good business to be in. Dad says it was the saddest thing he's ever seen and prays that this country will never go through anything like it again.

What still stands out in his mind was the number of poor people who didn't have anything.

"In our area, the worst was when the automobile plants shut down," he says. "A lot of people had come up from places like southern Illinois hoping to find jobs in factories. When they shut down, they had to give up and go home."

Peddling Potatoes

Dad remembers when they couldn't sell any farm crops except potatoes. "I remember driving around trying to get the best price for potatoes and settling for 35 cents a bushel," he says.

"I hauled them right into the store and put them on the rack. We were happy to sell them—that's how tough it was. There was no market for anything."

COW COMFORT. As the dairy herd grew from 20 cows in stanchions toward 50 cows milked in the parlor, more of the old barns were used for housing.

In some parts of the midwest, corn went for as little as 8 cents a bushel. A county courthouse in Iowa even burned corn for heat because it was so much cheaper than coal.

Regardless of how bad the farm situation was, it was far worse elsewhere. Millions of people were hungry and naked, while food and fiber were turned

"Farmers were better off during the depression than folks living in the cities..."

back or fields were left untilled.

In 1932, more than 6 million hogs were killed—not for food, but because farmers couldn't feed them anymore. Less than 10 percent were saved as food and distributed in relief, while most were used for fertilizer.

"Farmers were better off than people in the cities. No one had any money, but at least the farmers had shelter and could put their own food on the table," Dad remembers.

Disease And Hunger

Many people died of disease resulting from malnutrition. If they were lucky, welfare agencies or religious missions set up food lines. These free meals consisted of watered-down soup, beans or stew, but they undoubtedly kept thousands of people from starving.

In our area, General Motors contributed money to help and dispense food. The Oakland County Ford Division also provided needed food and clothes. When relief wasn't available, people had to steal, beg or eat scraps from trash cans in order to survive.

Thousands lost their homes and couldn't pay the mortgage. In 1932, more than 25,000 fam-

EXPENSIVE MILK CANS. If we had $50 for every leaking milkcan ever tossed in the dump, we'd be rich. Yet that's what many city folks are willing to pay to decorate their homes with these old milkcans purchased at flea markets and in gift shops.

ilies wandered through the country seeking food, clothing, shelter and jobs. Youths traveled in freight trains and lived in "hobo jungles" near train yards. People all over the U.S. had to find "shelter" in rusted-out car bodies and shacks built from flattened tin cans and old crates.

No Money

"During the depression, we were fortunate in that we always had food but no money," says Dad. "We never went without food or clothes.

"But my dad had a string of bad luck and lost all his money in the bank, along with the other investments he had."

At the outset of the depression, the road tax just about put Grandpa Frank under. In those days when a road was built near your home, property owners were expected to pay for it.

There was a time when he was paying for three roads at once—Baldwin Road, Clarkston Road and Indianwood Road. Because the farm was on Baldwin and within a mile of Clarkston and Indianwood, he had to also pay on those other two roads.

Down The Drain

Grandpa Frank was also a stockholder of the State Bank of Orion at the time it closed. He had double liability, which meant he had to pay back twice as much money as he owned in stock. He lost several thousand dollars.

During the early 1930s, banks were failing everywhere and people couldn't withdraw their money—money they worked their entire lives to save. The president of the nearby Pontiac Commonwealth Bank, who faced double liability like my grandfather, couldn't take it anymore. He shot himself.

Grandpa also owned some lots in a cemetery and lost all the money he put into that. Included among his many different investments were a clay products company, an insurance company

---

"After only 3 weeks of working in the factory, Dad was back on the farm..."

---

and Detroit Packing Company. He lost money in all of them.

When he died in 1949, nearly all his money had come from doing hard work—farming.

Back To The Farm

In the mid-30s, Dad heard about electrician licensing, so he studied for the exam and passed. He began doing electrical work while still working on the farm and then got involved as a full time electrical contractor.

He was working as an electrician when World War II broke out. Because of the war, there was a supply shortage (mainly copper needed for wiring), so Dad took a defense job at Pontiac Motors working on a Bofor gun.

"After just 3 weeks at the factory, my dad was stricken with heart trouble, so I had to go back and take over the management of the farm," he says. "We gradually sold off all the grade dairy cows and started a purebred Holstein herd."

Things were tough when I was about 10 years old.

Tough In The 50s

I remember once when Dad and I drove over to Birmingham for a job interview at Ford Tractor's experimental plant. The farm wasn't making any money and we needed another source of income. It turned out he was offered a job working as an engineer on a hay baling research crew in Arizona, but would have to be away from the farm for as long as 6 months.

Because of the financial situation, he and Mom had to think about it, but he declined the offer because it would be too long to be away from Mom and my three-year-old sister, Janet.

Pouring Milk Out

We were also lucky to have a good working relationship with Lakefield Farms, who began buying our milk in 1934. We didn't have to worry about any deductions—not even trucking costs. They picked up the cans of milk and paid us the going market price for the milk. After processing, they sold the milk to both the J.L. Hudson department store and Harper Hospital in Detroit.

In 1956, they went out of business and we began selling our milk to a creamery in Pontiac. All of a sudden we were being charged for everything, including transportation, insurance, bacteria discounts and many other costs. We had never dealt with any of that since we started in the dairy business in 1923.

When I was growing up, there were several years when the farm didn't make any money. During the 1956 milk strike,

HEIFERS DOWN, HAY UP. The wooden trusses for this barn arrived at the Lake Orion lumberyard in a freight car.

DARK DAYS, BRIGHT DAYS. Even though times could certainly be tough on the family farm, there were always many very enjoyable aspects to farm life. Watching the sun set on a frosty afternoon through the old wire fence could bring smiles to the faces of family members.

dairy farmers dumped milk rather than sell it for only $8.00 per 100 lbs. Farmers in nearby Oxford were dumping their milk down the sewers.

At one point, we had to let the hired man go, since we couldn't afford to pay him. Dad had to work a lot harder. I also was very involved in the farm—and was paid an allowance of $10 per week when in high school—a great deal of money in those days. But that wasn't so bad since I replaced the hired man.

I still remember when I was a senior in high school and my mother's aunt died. Mom and dad left for the funeral in Canada and with no hired man, they put me in charge of everything.

For 4 or 5 days, I handled everything—all the milking and chores—and went to my classes. It was a challenging time for me and taught me a lot about the responsibilities of farming.

Another Change

When labor started getting scarce and milk prices dropped sharply in the late 1950s, Dad decided to get out of the dairy business. The herd was gone by 1958 and replaced with 2,000 laying hens.

We sold the eggs to local schools, stores, summer camps and housewives. The poultry and cropping operations could be handled without hiring any labor outside the family.

Poultry wasn't the answer to all our problems, however. We ended up being tied down even more, since we had to gather eggs four or five times a day.

Dad became township supervisor in 1961, but kept farming a little on the side. He leased the ground out for a couple of years, but that didn't work and he sold some of the equipment.

Lohill Farm made it through America's darkest days and Dad says no one can truly understand what it was like except for those who had to endure it. And Dad, like every other person who survived the Great Depression, will never, ever forget it.

Five Generations Have Rocked Here

ONE SPRING, our 11-month-old granddaughter, Casey, rocked for the first time in our farm family's miniature antique rocking chair at my sister's home.

Casey represented the fifteenth member of our family stretching back to 1873 who has rocked in this family heirloom.

Who can count the times when five generations of the Lessiter family have rocked in this chair? There have also been countless numbers of friends, neighbors and other relatives who have enjoyed the chair which has a special place of honor in my sister's living room.

Grandma Was Only 3

The miniature rocking chair was originally purchased in Oxford for my Grandmother Norah when she was only 3 years old. Her father had driven his team of horses on the 14 mile roundtrip journey from the Seymour Lake farm where they lived to Oxford on one of his weekly trips to town for feed and supplies.

When he arrived back at the farm, the two ladies of the family were very excited since there were two rocking chairs sitting there in the wagon.

The miniature rocker was for his 3-year-old daughter. And the full-size matching rocker was a gift for her mother.

While the miniature rocking chair is still going strong, the full-size rocker didn't quite make it as far along in years—although it had more than 90 years of age on it before having to be replaced.

The chair was eventually handed down from my great grandmother to my grandmother. But that really didn't take place until 1945 since my great grandmother lived 3 months past her 100th birthday.

Eventually, Dad inherited the old rocker and my Mother kept it in our kitchen where it was used day-after-day for many years. It was used to rock four generations of kids and grown-ups alike over the years before it finally wore out.

A newer, yet similar rocker took its place. Both rockers of-

FIVE GENERATIONS have rocked here since this miniature rocker was first brought home to the Seymour Lake farm in 1873.

THE FIFTEENTH CHILD. One-year-old Casey Grabow represents the fifth generation of Lessiter family members to enjoy this miniature rocker.

some important food items in there that she didn't want either me or my sister to get into without her permission.

This rocker was a long-time favorite of not only our own family, but many friends who visited—especially on a cold winter day. The old rocker sat in front of a hot air register in the kitchen so you could get warm by sitting in the rocker after coming in out of the cold.

Countless friends also rocked here, along with a busload of foreign students from Wayne State University in Detroit who came out once to tour the farm.

Most of the group took a turn rocking in the chair. They expressed surprise and approval at finding this comfortable old rocker in an otherwise very modern farm kitchen.

Many Memories

Yes, the little rocker brings back many fond memories. Some 15 members of the Lessiter family through five generations have rocked contentedly in it along with many babies, their dolls and teddy bears who have been rocked to sleep here.

Plus, many daydreams came about due to the accompaniament of the back and forth rocking motion of this chair.

Despite what it lacked in size, its high back and sturdy arms always seemed to give even the smallest child a deep sense of security and Lessiter family love over the past 120 plus years.

fered plenty of comfort to members of our family and they occupied a special spot between the stove and a set of cupboards.

Chocolate Chips, Rockin'

Actually, those cupboards were a pain to get anything out of since you had to pull the rocking chair out into the center of the kitchen to get them open. So there was never a lot of really important food items kept in these cupboards.

But I do remember it was possible to open the cupboard door about 2 inches without pulling the chair out. That was enough space to reach in and grab a couple of handfuls of chocolate chips when Mom wasn't looking. And I think she did store

When Stagecoaches Brought The Mail

FROM 1854 UNTIL 1887, the western Lake Orion area's post office was located in a room of the original farm house at Lohill Farm. That room later became part of the farm's tenant house which eventually served as my sister's family home for a number of years.

Known as the Jersey Post Office, the outlying office was officially established by the federal government in 1847. Our family took over operation in 1854 of this official old-style Michigan post office and moved it into the little bedroom in the old farm house.

My Great-Grandfather John Lessiter, who came from England in the late 1840s, served as U.S. Postmaster for a total of 33 years. Later, my Grandfather Frank delivered mail during the 1920s along some of the early rural free delivery routes which had previously been established throughout the area.

New Jersey Connection

The Jersey Post Office actually got its name from the large number of settlers living in the immediate area who had earlier migrated from New Jersey. The mail used to arrive on the stagecoaches running along Baldwin Road on the 30-mile Pontiac to Lapeer route.

All official U.S. government postal business was transacted from a postal table passed down through the years among generations of our family. Along the back of the table, a rack of pigeon hole boxes held mail for each of the area farm families.

The ledger where postal transactions were recorded still survives with notations on the number of stamps purchased each day.

The Postmaster's pay wasn't too high by today's standards. Old records show my Great-Grandfather was paid $1.50 per quarter or $6 per year for serving as the official government Postmaster along with handling his farming duties. One fringe benefit of being Postmaster was conversing with friends as they picked up their mail and thus knowing everything going on in the area.

Two In, One Out!

Yet there wasn't a whole lot of mail to handle either. In one quarter of 1863, there were only 12 pieces of outgoing mail and 24 pieces of incoming mail. That

ALL OF THE OFFICIAL Jersey Post Office business was done during the mid 1800s from this table situated in one of the rooms in the original farmhouse. Missing today from the rear of the table is an upright rack of numerous pigeon hole boxes which held mail for each of the area's farm families.

sounds like the old-timers in this area liked to read more than they liked to write!

But with the absence of radio, movies, television and convenient transportation, getting mail was much more important in those days. This was long before the days when home delivery of mail got its start.

If you lived in town, you might walk over to the post office to check once or twice a week for mail in your box. However, country folks had to travel to the nearest office normally located along a stagecoach route or railroad line. This often meant having to hitch up a team of horses to go get the mail. And having a post office located closer to where you lived, such as the Jersey Post Office, made it much more convenient to get your mail.

Railroad Meant Changes

In 1887, the post office was moved 4 miles south to Cole Station which is now known as the Randall Beach area of the township. This resulted from the building of the railroad through that area and signaled the end of mail arriving by stagecoach.

With rails being laid closer to Lake Orion and Oxford and spreading like spider webs, many new postal stations were quickly established throughout the township.

The Place To Be!

But for 33 years in the mid 1800s, everyone living in the area between Lake Orion and Clarkston came to the Lessiter farm home to pick up their mail and to learn the latest news concerning friends, family and happenings all around the world.

Lohill—Farm, Drugstore And Even A Grocery

LIKE MANY OTHER farm families in the good ol' days, we didn't always have the luxury of doctors or grandiose supermarkets out in the country.

Back then, you relied on one doctor for medicine and a general store in town to buy your food. And when money was short, neither one was a very attractive option.

Home Remedies

Home remedies were very common among farm people. The distance from town and the horse and buggy doctor helped promote this practice.

Dad still remembers all the home remedies used by the family—which strangely as it may sound—often worked like a charm. And after looking at some of these "prescriptions,"

CABBAGE NO DELICACY. Dad was never fond of cabbage, particularly when it was cooked. His lack of interest goes back to the days when they used to ferment cabbage in a big crock for making sauerkraut.

BIG ON GARDENING. Like most farm families, the Lessiter clan raised much of their food in giant-sized gardens. While Dad's row markers were homemade, farm-sized implements were used to work the garden and to keep it cultivated during the summer.

you might agree with Benjamin Franklin's thoughts on professional medicine: "God heals and the doctor takes the fee."

Among those remedies most commonly used in the Lessiter household was a unusual salve made from pine pitch, beef bile (picked up from the slaughterhouse) and a number of other undisclosed ingredients. Applied with a stick, it was used to treat scratches, infections or just about anything.

"It's healing power was miraculous," Dad says. "It did the trick when all else failed.

"You would set it on fire, dab it on a cloth—and while it was still hot—you'd slap it right on the sore and it would clean the infection right up. I think putting it on there hot probably did as much good as anything."

Grandma Norah used to have a formula for linament she gave to the druggist to make up for her. I don't remember exactly what it was, but it had wintergreen in it—and it really smelled good.

Dad recalls the time when he used it on Mom. "Being a city girl, she wasn't too happy with the concoction," he says. "It practically took the hide off her back."

Watkins' linament was another favorite used externally for all sorts of muscular and rheumatic ailments. Sometimes, it was watered down and taken internally for other complaints.

To relieve stomach aches, they used to give family members whiskey and sugar—and Dad says it sure did work. Added to hot water, it was also administered to babies suffering from colic.

Medicine Worth Taking

A hot whiskey toddy was often a sure cure for a cold. A mixture of honey and brown

---

"When you were sick, a hot whiskey sling could sure make you sweat..."

---

sugar was given to relieve a cough.

Dad says they never made their own whiskey, but "a hot whiskey sling sure would make you sweat."

In addition, blackberry liquor was effective for stopping diarrhea. Castor oil or epsom salts were used for the opposite effect. For a chronic case of constipation, senna leaves were either ground to be eaten or boiled for tea.

Mustard And Lard?

More serious chest colds and pneumonia were treated with mustard plasters or onion poultices. Remembering the burning it caused, Dad says it generated enough heat to kill anything. Turpentine and lard was another application for the chest and neck to relieve bronchial distress. Goose grease was still another cure.

For bee stings, they went out to the swamp, dug up some mud and packed it on the inflamed area. Vinegar, butter or a pack of tea leaves were effective for minor burns.

Earful Of Smoke

Dad remembers a common earache remedy used when heated drops of oil or glycerin didn't work. His Grandfather would smoke a pipe and then blow the warm smoke in the affected ear. For toothaches, they applied clove oil—or a hot pack.

Molasses and sulfur were administered every spring as a tonic to cleanse the system of the winter's ills. People with head lice were doused with kerosene.

Dad recalls every home had a large bottle of Lydia Pinkham and Father John's Medicine as "cure-alls" for just about every ailment.

A copy of *Dr. Chase's Doctor Book* belonged in every farm home—second in importance only to the Bible. He remembers his Mother and Grandma paging through it often and following its advice closely.

Usually in every neighborhood, there was a woman who reportedly possessed considerable knowledge of various herbs used to cure various ailments. As a last resort, farmers called the family doctor and the sight of his "rig" told the neighborhood someone was seriously ill or a baby was being delivered.

Self-Sufficient In Food

Dad remembers when they used to keep cabbage in a barrel in the garden or in a big flower crock out by the milkhouse. When asked about that cabbage, he cringed. "I never liked that cabbage coming out of a barrel," he says. "It used to come out rotten and slimy."

They made sauerkraut every fall. Our family liked to do everything on a big-scale, and once made 20-gallons of sauerkraut—it all spoiled.

They dried apples, potatoes and other vegetables and fruit

> *"If you wanted a chicken for Sunday dinner, you just chased one down..."*

from the garden. They were stored through the winter and a fire was needed in the cellar to keep them from freezing when it was zero weather outside. Grandma Norah dried slabs of meat to make corned beef.

Grinding Breakfast

Back in the depression, Dad used to grind wheat in the feed grinder in the barn for breakfast cereal and graham flour. Years ago, there was an old one-cylinder gasoline engine to power the feed grinder.

Dad says the family also used to butcher their own hogs for pork chops and hams. They also used to store quarters of beef they bought at the locker plant in town.

On Sunday morning, if you wanted a chicken for dinner, you just went out and chased down a chicken. "We'd kill it, pick the feathers and cook it right away," he adds. Talk about fresh!

Dangers With Cider

Dad remembers several times how cider stored in the basement got the best of him.

"When I was about 10, I went down to get a drink of cider," he says. "It was so sour that I put sugar in it and I guess it turned to alcohol. I got so drunk I put my dad's rubber boots on over my shoes and couldn't get the darn things off.

"After drinking that hard cider, I came up to the dining

HARVEST TIME. Eating hand-picked, vine-ripened vegetables from the garden in late summer and fall was always among the real benefits of growing up on the family farm.

OLD HOG COOKER. When butchering hogs for family use, water was heated in this old concrete tank which featured a wood burning unit to heat the water to boiling. The tank was located in the back of the farm shop and even had its own brick chimney.

room to unlace my shoes. They say the next thing they knew, I was laying on the floor."

Come A Long Way

Right up until the time Mom died, she and Dad still canned and preserved a great deal of food from their huge garden. They found something special in sitting down to a good, home-cooked meal, with food taken right out of Lohill Farm's soil.

At any rate, both medicine and food preparation have come a long way since those days—from over the counter pills for every problem to frozen pizzas and TV dinners.

But people who used those remedies truly believed they worked. I ran into a nurse over at our local hospital yesterday and told her about some of the Lessiter home remedies. She laughed at some of the treatments once used by families all over the U.S.

There was some "medical benefit" to a few, however. She said packing mud on a bee sting provided the same type of topical anesthetic effect she would recommend in using baking soda today. She also noted mustard plasters and onion poultices might keep the patient warm. Although kerosene would definitely kill head lice, you'd never catch her advising it.

Her overall assessment of rural home remedies: "Well, we certainly don't prescribe that sort of treatment today."

Nevertheless, whether it was psychological or medical, it got our family and others through some tough times.

"God heals and the doctor takes the fee."

The Old Tire Swing

IT NEVER FAILED. Whenever we made the 325-mile trip back to the farm from our home in Wisconsin, our four kids would pile out of the car and immediately greet their grandparents, uncle, aunt and cousins.

As we arrived after the long drive, Grandma would usually have milk and cookies or other snacks prepared for the kids. But just as soon as the snacks had been finished, the adults would stop talking, look around and realize all seven grandchildren were gone—they had already headed for the barns.

The first item on our kids' list

BIG OLD MAPLE. Anchoring the tire swing, this maple was planted in the center of the farm yard a number of decades ago by Frank Lessiter. It was among more than 100 maples planted around the Lohill farm buildings and along a mile of Baldwin Road.

THE DELIGHTFUL PEACE AND QUIET THAT ONLY A VISIT TO A FARM CAN OFFER. Susie Lessiter enjoys some quiet early morning time alone on the tire swing without the noise and hassle of her brother, sisters and cousins.

of "must things to do at the farm" always seemed to be taking a ride through the air on the old tire swing.

It didn't matter whether we made the trip back to Lohill Farm in good weather, rain, snow or sleet—the kids were always anxious to head toward the barn and the old tire swing. Except for those times, I guess, when we arrived after dark and they would have to wait until morning before journeying out to the barns.

Fly Through The Air

For many years, a rope and a long-discarded tire from one of the farm's trucks hung from a strong limb on one of the beautiful maple trees planted decades earlier in the yard by Grandpa Frank Lessiter.

The kids had their own swing set in their yard back in Wisconsin. But a ride on the old, cheap rusty store-bought swing could never come close to the thrill of a trip on the swing at the farm.

It was always fun to ride the swing and have somebody push the tire high into the air. There

"The old swing is a lasting memory of the many good times spent at the farm..."

was room on the tire for two or three kids at a time, which always made for extra special fun during Christmas-time visits when someone might get knocked off into a snowbank.

Even the three cousins who lived on the farm, and often took a turn each day on the swing, seemed to think it was a special treat to swing with the newly arrived grandchildren.

Not wanting to miss all the fun, they didn't think the Wisconsin kids should get extra turns just because they only came to the Lessiter family farm a couple of times a year.

As you might guess, plenty of arguments were soon going on between the grandchildren as to who would get the next turn on the swing.

The Rush To The Swing

When things quieted down later in the visits to the farm, each of the kids would often be found at the swing all alone. With a quick push of their feet, they would be gliding through the air and enjoying the peace and quiet which only the farm could bring.

Or maybe one of the friendly farm dogs would be snapping at a pantleg as the youngster's feet

SAILING THROUGH THE AIR. Susie Lessiter fully enjoys the ride.

CHRISTMAS ON THE FARM. During those visits home for the holidays, the kids would spend many enjoyable hours playing with cousins on the old swing.

glided by on the swinging tire.

The old tire swing is a long-lasting memory for all seven Lessiter family grandchildren—regardless of whether they lived on the farm or 325 miles away in Wisconsin.

Saturday Nights On The Farm

Dad fondly remembers what Saturday nights were like on the farm back when he was growing up in the 1920s.

When all of the farm chores were finally done on Saturday afternoon, the family would pile into the car for a trip to town. This would lead to an exciting night with other folks from the country.

"Your grandmother always had plenty of time for Saturday night shopping in Lake Orion because my Dad would play cards with his friends in the back of the pool hall for a few hours," recalls Dad. "Sometimes he would play to 11 p.m., leaving your grandmother and me waiting for him in the car.

"We sometimes went to a movie or would walk over to the amusement park on Park Island and take in a few rides while he was playing cards. Sometimes we would wait quite a few hours for him.

"But your grandmother would never go in and haul him out of the pool hall—preferring to just sit in the car and wait for him."

Farm Mothers Are One Of A Kind

AS YOU GROW older, you learn there aren't many things in life that you'll have forever. As your life takes all those unexpected twists and turns, you always risk losing many things—your farm, your house, your loved ones—but one of the things you truly own are your fond memories of how things used to be.

Without a doubt, growing up as a youngster on Lohill Farm provided me with some of the happiest, most carefree days of my life. I live far away from that farm today, but I still recall spending those days working and playing among our rolling green hills. And synonymous with every one of those memories is my Mother.

There's a certain relationship between a mother and her son or daughter that is shared long after the passing of generations. When some people remember loved ones who've passed away, tears well up in their eyes. It was the spring of 1982 when my Mother passed away, but still, when I think of my Mother, I can't help but find a way to smile.

A Mother Like No Other

Born in Canada, Mom was a city girl who had never lived on

A PROUD GRANDMA. Mom very much enjoyed the first of her seven grandchildren, Debbie.

TRELLIS OF ROSES. Every spring, this beautiful display of pink roses would blossom on the back of the garage. Mom and Janet would always take time to admire the latest display of beautiful blooms each May.

a farm until she married my father. I know it was quite a change of pace for her, but she soon realized the advantages of country life. Although she never did get the courage to lead a horse while cultivating her garden, she loved that farm and was as much a part of its operation as Dad, the hired hand or anyone else.

She was the happiest person I ever met and even when she was fighting for her life, she still managed to smile when I walked into the hospital room.

I could go on for pages, but here are just a few of my favorite memories of Mom.

She was known throughout the county for her special chocolate chip cookies (see page

"No matter how bad things might get, her magical baking touch would always help..."

352). No matter how bad things became at school, work or in any other aspect of my life, her magical baking touch could always bring me up again.

One time I became so swamped in schoolwork my last year in college, I was forced to stay at school over the Thanksgiving holiday. It was one of the loneliest days I've ever spent, but sure enough, Mom was knocking at my dorm room door the day after Thanksgiving, having driven 4 hours so we could have a special Thanksgiving of our own.

She could be tough too. When I was in the eighth grade, I refused to memorize the poem, "Abo Ben Adem." When the final schoolbell of the year rang in mid-June, I set my sights on spending a week with my friends at Michigan State University for

4-H Club Week.

But when Mom saw my report card and the note from the teacher explaining how I had stubbornly refused to learn the poem, she put her foot down. An English teacher by trade, she in-

"I finally learned the poem, but only because she put her foot down..."

structed me that I wouldn't be making the 4-H trip until I memorized the poem. I still didn't want to do it and I didn't really think she'd make me learn it.

After 3 weeks of arguing, I realized she was serious. I learned the 50-line poem and recited it for her in the living room 2 days before the bus left. I still remember it to this day. If I'd learned it 6 weeks earlier, I could have earned an "A" instead of a "C" in the eighth grade English course.

Another time, I'd returned home from college for the summer and was working on the farm.

Shave Beard Or Don't Eat

Against her wishes, I began growing a beard. She had no luck getting me to shave until one day at noon when Dad and I came in from the fields for dinner only to find no food on the table. "Where's dinner?" Dad asked.

She answered, "I'm not mak-

ALWAYS A HAPPY COUPLE. For 44 years of married life, Mom and Dad were a very supportive couple and always enjoyed life, despite some health problems which Mom suffered through.

A REALLY GREAT LADY. Mom was the most caring person our family members have ever had the opportunity to know. Most importantly, she influenced many, many lives, both inside and outside of the Lessiter family.

MOTHER, DAUGHTER. Mom and Janet always had a very special relationship and enjoyed doing many things together. In later years, Janet and her family lived 100 yards up the road on the home farm. This made for a great day-by-day relationship between Mom, her daughter and three delightful grandchildren.

ing anything to eat until Frank shaves his beard."

I got up from the table, went upstairs and shaved and we had

Pleasant Memories...

ALL OF US sitting here this morning are sharing a great sense of loss. Here with us in spirit this morning, I know my Mother would like to have each of you remember all of the pleasant memories you had of her.

Many of you here today go back much farther with Mom than do my sister, my uncle, my father or me. You can remember the high school days, the women's softball teams, the swimming races in Lake Orion and many other wonderful times.

In those early days, she adopted the philosophy that was to always be an important part of her life. It was this: *People are always more important than things ever are.*

She Always Helped Others

She taught all of us to take time to do things for people who are important in our lives before it is too late...making a telephone call, paying someone a visit or taking 5 minutes to write a brief note on a card. Nothing was ever more important to my Mother than her family and her many friends.

Importantly, she never judged people. She was able to accept them for what they were, always accepting their faults without trying to change them. She challenged people to do their best in everything they did and encouraged them to turn to their faith for strength.

Now, I'd like to share a few personal memories that members of our family have of my Mother.

First of all, she and my father led a fantastic 44 years of married life. You could never find two people more in love than my parents...and they set a remarkable example which the rest of us family members have always tried to follow.

She Was A Big Fan Of The Lake

Many people have told me in recent days how as teenagers Mom, Aunt Margaret and Uncle Sherm practically lived at the lake the year around—swimming all summer and skating all winter.

Our own children were always asking Grandma to tell them once again how she used to swim in races all the way to and from Park Island from Green's Park in Lake Orion. I still remember the first time Grandma took our kids to the lake and showed them how far it really was—they were shocked.

Over the years, we too had concerns like other families. But Mom was always there to provide needed strength to work out any problem.

When my sister used to come home from school on a rainy day, Mom used to lovingly hold her tight in the kitchen rocker and brighten her mood.

Among her proudest accomplishments were the two 4-H Parliamentary Procedure teams she coached to state championships. She also tutored students over the years. She loved a challenge and

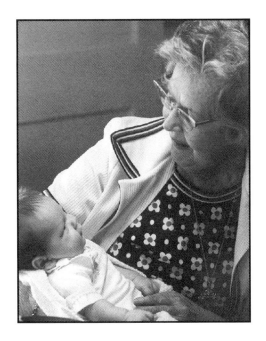

ANDY AND GRANDMA. With Janet and Neil both teaching school, Grandma did a considerable amount of babysitting for her three grandchildren, including Andy.

a good, but late, dinner that day!

I still remember how thrilled my own children were the day she surprised them for "Grandparents Day" at their grade school. Because both sets of their grandparents lived out of state, they were a little disappointed when my wife and I told them Grandma and Grandpa couldn't come to their school for the day's events.

But our second-grade son didn't know any better and wrote

> *"Mom always did everything she could to help make everyone happy..."*

her anyway. As a result, she flew over and surprised the kids at school. Only a special Grandma

A VERY SPECIAL VISIT. Grandma Lessiter's first grandchild, Debbie, shows two of her own children, Alex and Molly, where their great grandmother is buried. Over the years, Debbie has passed along to her children many of the beautiful experiences which she enjoyed along with the Lessiter family stories told by her grandmother. Looking on are Grandpa and Pam.

always loved to help others who were in need.

I'm always reminded of the time she went to Europe with my Dad and sister. Quite frankly, she didn't want to go. But she knew Dad had his heart set on making the trip and she did it for him.

She Gave Europe A Shot

Once she went, she had the time of her life. And one of the very special moments came when more than 200 people sang "Happy Birthday" to her in a Lucerne, Switzerland, restaurant, including many Swiss who hardly knew a word of English.

Her seven grandchildren will cherish forever the special stories she wrote...stories that included each of them interwound right into the story's plot.

Over the years, my Mother directed many plays and ceremonies. And if she was in charge here today, she'd want to wrap this up on a happy note. So I'll do my best to tell two little stories I think she'd really appreciate.

The Christmas Card Caper

A few years back she was at our house and was looking through our stack of Christmas cards. That year, she and Dad had sent out a bright red card with a picture of the old barn on it. As she came to that one in our pile of cards, she opened it and it read, "From Donalda and John."

She laughed and said, "Well, somebody must have certainly been surprised when they opened our card and it said, 'Love, Mom and Dad'."

Then there were her delicious chocolate chip cookies. Many of you here have probably baked her special chocolate chip cookie recipe over the years, as have my sister, Janet and my wife, Pam. But somehow, nobody can ever take that recipe and turn out the same kind of delightful cookies I remember my Mother baking.

She enjoyed many laughs over the years with Pam and Janet about her recipe. But neither has ever been able to match the cookies Mom baked.

Mom was one great lady...and her spirit will continue to be with all of us for ever. She influenced a great many lives. But best of all, she was our Mom and we loved her very much.

—*April 20, 1982*

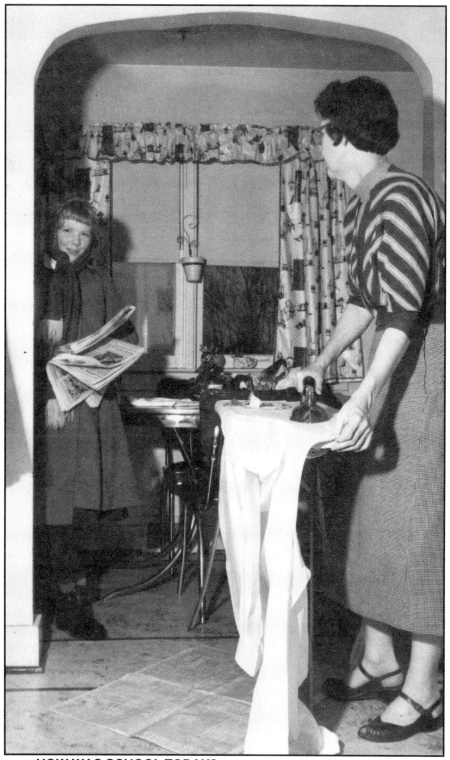

HOW WAS SCHOOL TODAY? Mom was always glad to see Janet as she brought in the newspaper after getting off the elementary school bus. She would always have a snack waiting for her favorite little girl.

a piece of cake waiting for you as you walked in the door from school.

She was the one who made you chicken soup and read you books while you were sick. During tougher times, Mom wore the same old dress for another year so you could have that train set or brand new doll you wanted for Christmas or your birthday.

After you grew up and moved away, Mom wrote you a letter once a week letting you know how much she loved and missed you. When you returned home,

> *"Spending a few minutes remembering Mom always brings a smile to my face..."*

she was the happiest of all, even when you greeted her with a bulging bag of dirty laundry.

You could always come home anytime and Mom made you feel like you had never left in the first place. She sat for hours with her grandchildren on each knee, telling the wide-eyed youngsters what their Mom or Dad was like when they were growing up.

It's hard to ever realize the great influence your mother had on your life. You grow up thinking you're your own man or woman, but when you think about the person you turned out to be, you can't begin to think where'd you be today without her.

If you're lucky enough to be able to, give your mother a call soon and thank her for making you the person you are today. If you can't, spend a few moments remembering...it's sure to bring a smile to your face.

would do something like that.

A Day For All Moms

Think back yourself. Ever since you can remember, your Mom probably did everything she could to make sure you were happy. She was the one who had

When Dad Really Got Sick

IT'S BEEN more than 50 years, but I still fondly remember the first vacation our family took on the shores of Lake Michigan.

After running barefoot over the dry dirt of our family's farm fields, playing in the sand and water of this mammoth lake was a memorable experience for this 5-year-old. But it wasn't such a great time for Dad.

The doctor ordered Dad to take that vacation due to "dairyman's disease," a problem once fairly common to many livestock farmers.

"Dairyman's disease" is a chronic infection known as undulant fever in humans and brucellosis in cattle. It was one of the most serious and widespread diseases affecting the livestock industry in those days, so thank goodness brucellosis is no longer the problem it once was.

Bad Deal! Bad Debt!

Dad came down with undulant fever in May of 1945 after doctoring a sick cow my grandfather Frank had taken to settle a debt owed by a neighbor. As Dad recalls, "Everyone would have been way ahead if we'd just forgotten about that darned debt."

One of the first disease symptoms Dad had was a troublesome sciatic nerve in one of his legs.

BEFORE TREATMENT, Dad was in bad shape from the undulant fever as shown in this photo with Frank and the farm dog, Scotty.

He couldn't stand up straight and was soon bent over constantly with pain. At least once, he crawled from the barn to the house. Then came chills, headaches, nighttime sweats and fever followed by a plain overall weakness.

The real concern was nobody in the medical community seemed to know what the problem really was.

"I tried chiropractic treatment—that didn't help," Dad recalls. "A local doctor prescribed large doses of vitamin B—that didn't help either.

"I went to the University of Michigan hospital clinic—even those doctors couldn't find what was wrong with me.

"Then somebody recom-

DURING TREATMENT with a revolutionary antibiotic and enjoying a rest with the family along Lake Michigan's shores, Dad got much better.

mended a doctor in Detroit. He took all kinds of blood tests, then finally diagnosed undulant fever."

Cure Vs. Disease

This doctor treated the problem with what was a newfangled drug in those days...Aureomycin. The doc later told Mom this was the first time he'd ever prescribed this new antibiotic—one that has since enjoyed a long history of success.

Returning home, Dad started taking those Aureomycin pills. Then he got a terrible reaction to the medicine.

"His heart was beating like mad and he was very weak from taking the medication," remembered Mom. "I called the doctor and told him about the terrible reaction. He told me to immediately drive John back to Detroit—a tough 2-1/2 hour trip.

"By the time we got to the doctor's office, your Dad was nearly in a coma. Within minutes, the doctor gave him an electrocardiogram. I was really scared."

As it turned out, the doctor had given him too large a dose of the new antibiotic. In fact, Dad should have been hospitalized while the doctor experimented with the proper dosage.

Miraculously, Dad somehow survived 3 weeks of treatment.

"Besides coming close to death, those pills cost 60 cents each and I was taking a half dozen of them each day," he says. "That $3.60 per day was a lot to spend on medicine in those days.

"Worse, I darned near died when that doctor put me on the Aureomycin. But he was the only doctor who diagnosed my problem as undulant fever.

"While the treatment was rough, it cured the disease and I've never had any problems since."

Leave The Farm!

The doctor suggested Dad get away from the farm for some well-deserved rest and relaxation. So the three of us went to Ludington, along the eastern shore of Lake Michigan—a rest for Dad, but the first "vacation" away from the farm for me.

Even more than 50 years later, I still remember that week as a glorious time of playing in the sand and water. We stayed in a rooming house along the canal where the Chesapeake & Ohio ferries docked to unload railroad cars, automobiles and tourists in those days.

It was great fun watching the Great Lakes boats come in, eating out, playing on the beach and sleeping in a strange bed.

One reason I remember the first vacation so well is that I've gone back to Ludington several times to relive those early day memories. Several times in later years, I took our six-member family on the ferry boat across the lake which ran for years from Milwaukee to Ludington.

The rooming house still stands along the canal and it was great fun pointing out where Dad, Mom and I had stayed some 50 plus years earlier.

Some events linger in the minds of farm kids for a long time. This is one of them for me.

I've always been grateful that this enjoyable vacation also meant Dad was cured of the dreadful "dairyman's disease."

Where's Grandpa? He's Fishing!

EVEN THOUGH he died when I was only 8-years-old, I distinctly remember spending many enjoyable days with my Grandfather Frank.

My Dad, John, and other family members used to tell me Grandpa Frank really loved fishing—even though he didn't go all that often.

While we owned our own 18 acre body of water called Dark Lake near the back of our farm, he preferred to fish in the bigger lake located a quarter mile behind the big farm house on the other side of the road. This is where he had thrown his fishing line many times as a child...and he definitely knew the favorite fishing holes well.

Actually, our family had ear-

A BIG STRING of tasty bluegills was a common occurrence when Grandpa Frank went fishing on the farm's two lakes.

AS A YOUNGSTER growing up on the family farm, it wasn't unusual for young Frank to accompany his grandfather on a late afternoon fishing jaunt to the farm's lakes. We often came back with a stringer of bluegills.

lier owned land along both the eastern and western shores of the 125-acre Dennis Lake which was later renamed Heather Lake. The new name came about as a tie-in with Heather Lake Estates which features executive houses built in what used to be corn, wheat, potato and hay fields which dotted the shoreline.

Back in 1923, my Grandfather and Grandmother sold 16 acres of hilly pasture on the west side of the lake plus another 16 acres of land under the water to Fred Zannoth. His real estate agent had pestered my grandfather that year during the Michigan State Fair in Detroit to sell this land.

After they brought the Shorthorn cattle home from the fair, they decided to sell the land to Zannoth. In recent years, 1 1/2-acre executive home lots sitting in these hilly pastures have been selling for as much as $165,000.

This land was later sold by Zannoth to Oscar Webber, who headed the big J.L. Hudson Co. department store in the heart of downtown Detroit.

Around 1926, Webber wanted the rest of our family's land which came up to the eastern shores of the lake. He was very persistent and finally made an offer my grandparents simply could not refuse for the remaining 55 acres, all shoreline rights and 32 acres of lake bottom land.

Dad says he couldn't recall anyone buying the ground under a lake before these two purchases. But Webber had made up his mind—he wanted to own the entire lake, regardless of what it took!

With this final purchase, Webber owned nearly 1,000 acres of land which went around the entire lake and also surrounded our farm on three sides.

However, our family was also persistent with some of the language in the sales contract. Part of the agreement in selling this acreage stipulated the Lessiter family could always fish there. As you might guess, that was an important part of the contract to my Grandfather!

Over the years, my Grandfather and Dad remarked many times that several of the fields which were sold contained the most productive soils on the entire farm.

Back To Fishing

Grandpa Frank used to tell family members there was always too much farm work which had to be done before anyone could go fishing. But if somebody showed up to go fishing, he never seemed too busy with the livestock chores or field work to drop a line and hook in the lake.

Dad can recall many folks who used to come out and go fishing with my Grandfather.

Grandpa's brother-in-law, Will Anderson, really liked to fish. For many years after his wife died, he would show up to go fishing without giving any advance notice.

"You never knew when he was coming or when he was going to go home," recalls Dad. "My Mother used to give him his meals and a room. He fished with my Dad or me and also went alone when there was too much farm work for the two of us to do.

"I remember ice fishing many times with him. He used to drive his old Dodge sedan right out on the frozen lake.

"Sometimes we would leave our fishing lines in the ice over-

Asleep At The Wheel!

Before I get into the details of one more lake story, I need to share a few facts with you about the small lake on our farm. Called Dark Lake, it covered 18 acres, was very cold and extremely deep.

In fact, the lake drops off to a depth of 60-feet just 3-feet off shore. In other parts of the lake, an electronic depth finder once marked the depth at 125-feet. "I can remember friends telling me they never had enough fishing line to touch the bottom of the lake," says Dad.

Corn and alfalfa fields ran to within a few feet of the shore...which is what this story is all about.

One afternoon after a hearty dinner, Dad was cultivating corn fairly close to the water's edge. With the combination of a big noon-time meal, plenty of hot sunshine and the boring cultivating task, he sometimes tended to doze off for a few seconds at the tractor wheel.

This happened one bright sunny afternoon right along the lake's shoreline.

Half asleep as he bumped over the end rows of corn at the lake's edge, he awoke to find himself and the tractor within 2-feet of taking a "header" into the lake. As you might guess, he hit both the clutch and brake on the tractor in a big, big hurry!

If he had dozed off for just a few more seconds, the cold water and the 60-foot deep plunge into the lake would certainly have brought him awake very quickly. And I don't know how we would have ever "fished" the tractor and cultivator out of the lake.

One thing was for sure—we never had to ever worry about Dad falling asleep again while cultivating corn near the lake!

WHERE THE KIDS NOW REST IS where the old livestock watering tank was located. Goldfish spent their winter vacations here and a supply of minnows were kept on hand in the old concrete tank for the occasional fishing trip.

night, then hurry down the next morning to take off the fish."

While he doesn't remember who it was, Dad says one of Grandpa Frank's friends once got caught by the game warden. He was either illegally spearing fish or got caught with a big basket of fish above the day's limit.

"We used to use gasoline-soaked torches to see and spear fish at night," says Dad. "The fish would just be laying there below the surface of the water. You could pull in a washtub full of fish in just a couple of hours."

Minnows In The Tanks

Dad says the family used to keep an ample supply of minnows on hand in the livestock watering tank out at the barn.

"We used to go out there, roll up our sleeves, take a net and try and catch some minnows to go fishing," he says. "It was darned cold in the winter."

As a kid, I remember goldfish were always swimming in the big concrete watering tank to keep algae under control and to keep the water from freezing over in the winter. These weren't your dime store variety goldfish either—many were 6- and 10-inches long. And fat!

My Grandmother also enjoyed a goldfish pond each summer in her yard and those fish would then spend the winter in the cattle watering tank.

Another dedicated fisherman who frequently showed up on the Lessiter family doorstep was one of the African-American doormen at the J.L. Hudson Co. department store in downtown Detroit, located 35 miles south of the farm.

Nobody seems to remember his name today, but from the late 1920s to the early 1940s, he was a frequent farm visitor on weekends or vacation time during the fishing season.

Oscar Webber, an executive at the department store and the owner of the neighboring Lakefield Farms, told the doorman to feel free to come out to the lake and fish.

"The only problem was that their farm didn't have a boat," says Dad. "So he always came to our place and used my Dad's boat to fish.

"For nearly 30 years, anytime my Dad, Mother or any other member of the family would go to Hudson's in Detroit to shop, we would always go in the Farmer Street entrance where this man worked.

"He would make a big fuss over all of us and could be counted on to recall all the great fishing experiences and the many big ones that got away out at the farm.

"He would get chairs for all of our family members to sit on while they waited for the rest of the family to finish shopping. He was always glad to see us."

In later years after my Grandfather passed away, I remember seeing the doorman many times when we shopped at the big department store. Long past-retirement age, he was always delighted to see our family—even though he hadn't dropped a fishing line in the lake in more than a dozen years.

One More Tale

Even with all of these great fishing experiences, there is one final story to tell about the 109 acres of lake property which the family sold in 1923 and 1926.

"When the family sold that last land and the lake access in 1926, they held on to family fishing rights to the lake," says Dad. "When the 1,000-acre farm was sold for real estate development in 1956, the lawyers asked my mother to sign off the family's fishing rights to the lake.

"She got $500 for just signing the paper which she thought was great. She always laughed about it and said that was the easiest money she had ever earned, since she never went fishing anyway!"

The Family's "Good Child"

AT MY DAUGHTER'S wedding, I overheard one of our other kids ask my younger sister, Janet, what it was like growing up with me.

Apparently, due to gross exaggerations by my Uncle Sherm, I seem to have developed the reputation of being the town demon as a kid.

After telling the crowd all she remembered about me was eating hundreds of chocolate chip cookies and wolfing down catsup sandwiches, Janet replied, "We fought all the time—he was always picking on me for one thing or another." Then she added sharply, *"I was the good child."*

So like all brothers, instead of setting the record straight among the wedding crowd, I decided to laugh it off and be mature.

My Reply?

Well, now, thousands of folks across the country will hear about what it was really like growing up with Sis.

Janet and I were the only two kids, and she's 7 years younger than I am. Unlike some other young girls in town, Janet loved romping around the farm from the time she took her first steps.

She had just about every pet imaginable and often joined right in with the farm work with Dad and me.

Later on with her husband and three children, she continued to live at Lohill in the first farm house that John and Nancy Lessiter built in 1854. It originally stood next to the big barn, but was later moved 150 yards south and has been added onto and renovated many times since.

Her Own Playground

Janet always had fun on the farm. She and one of the hired hand's daughters, Shirley Robertson, used to play "house" in the corn crib, creating their own benches, kitchens and living rooms.

She loved to walk down the

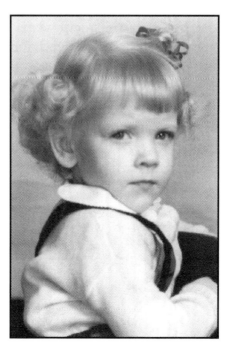

LOTS OF CURLS. Grandma Norah wanted to send this pre-school photo of Janet to numerous Hollywood film agents for consideration when they cast that next "leading lady."

THE LESSITER KIDS. At left, Frank and Janet are shown in one of those few moments when they weren't arguing with each other. Above, Janet poses in front of the fireplace at the Byron Chapin home during one of the Christmas parties always enjoyed by the two families.

lane to the lake and in the winter she'd skate around on one of the ponds with some of the other farm children.

She also loved petting the cats, which always irritated Dad. "Those aren't pets," he'd say. "They're just barn cats and you're going to get bitten."

Dad was right. One of those straggly old cats bit Janet once. We ended up having to quarantine the cat for 10 days to make sure it didn't have rabies.

A Little Mixed Up

When Janet was in first grade, her teacher, a bridge-playing friend of my folks, was teaching about various kinds of farm animals. Since Janet was the only one who lived on a farm, my parents invited all 22 first graders out to our place so the kids could see a working dairy farm first-hand.

Dad explained every aspect of farming to the curious youngsters and fielded all of their questions. To tie the whole experience together, the next day's art assignment was to draw what impressed them the most about what they'd seen on the farm.

While the favorite topic was the manure carrier track and manure pile, many of the children drew pictures of cows. After all

"She drew the udder on the front end of the cow..."

the artwork was turned in, the art teacher laughed when he realized one of the first graders put the udder on the front end of the cow—right under the cow's head.

But the real uproar in the faculty lounge came when it was discovered the artist was none other than the only farm-reared veteran in the class—my sister, Janet.

Mom Says "No Way!"

When something was going on out near the barn or anywhere else on the farm, Janet was always there. I remember one summer afternoon when "working with the guys" got her into hot water with Mom.

Janet was 6 years old and out by the barn one July day. As they worked, several of the hired hands took their shirts off in the sweltering heat. Seeing nothing wrong in doing the same, Janet removed her shirt and went about her chores.

Later that day, Mom stormed out to the barn yard and pulled a shocked young Janet aside. "Girls never take their shirts off," she said.

Bad Day For Janet

Janet and I differ on the interpretation of this next story, but

CLOWNING AROUND. The hired man's daughter, Shirley Robertson, and Janet get a few laughs out of lugging the milking machines to the barn. At right, Janet shows off her tractor driving techniques.

when I was 14 and Janet was 7, Mom and Dad left me in charge of babysitting Janet. Anyway, things got a little wild, and while I was dancing around on the parlor furniture, the glass shattered on my mom's prized coffee table.

Janet was worried about what Mom and Dad would do. I kept chanting, "Just wait till Mom and Dad get home, Janet. You're really going to get it—just wait."

After my relentless teasing, she got scared and started crying. But by that time, it was too late, and she got so upset that she threw up. Forty years later, she still hasn't forgotten that one.

As she got older, Janet was always helping with the farm work. She fed the calves, drove the tractors, fed the chickens and gathered eggs.

"I never wanted to work in the house, I always wanted to be out-

61

THE NEXT GENERATION. Janet helps Andy open one of his Christmas presents. At upper right, Janet helps oldest daughter Trista off the afternoon school bus, and goes over some school work with her as niece, Susie Lessiter, looks on.

side," Janet says. "And you can still tell that's true if you take a look at my house."

When she was about 10, I remember she helped out a lot driving the tractor during hay season. She had a terrible time backing up the trailer because she claimed she couldn't see.

Several times she'd run into something and the other workers and I would tease her. Finally, she's just stomp off and I'd hear about it later from Mom.

She got us back, though, even

> *"We would tease her, but somehow she always seemed to get even..."*

if she says it was an accident. We were in one of the hay fields near the back of the farm where there was a big swamp at the bottom of a steep hill. With me and a couple of other guys on the back loading bales on the hay wagon, she was having trouble controlling the tractor as she was driving down the slope.

Once we started giving her a hard time, she quickly popped the clutch and we fell off, nearly landing in the smelly swampwater.

Sure, it could happen to any-

YELLOWSTONE VISIT. As a teenager, Janet and the family made a late summer trip to pick up her brother after the summer he spent working at Carnation Farms in Washington. On the way back to Michigan, they spent several days sightseeing at Yellowstone Park.

TIME FOR RELAXING. Sisters-in-law Pam Lessiter and Janet Roberts enjoy well-deserved rest on the milkhouse steps. At right, the "family's good child" enjoys a good laugh.

one, but her roaring belly laughter made me think it didn't happen entirely by chance.

Hangin' Out At Lohill

As Janet got older, Lohill started drawing a largely female crowd. Because we always needed extra help in the summers, Dad always hired some high school boys to pick up some of the extra work.

When Janet's friends entered their teen years, she and her friends spent a lot of time out there by the barns. But as she'll admit, they weren't exactly lending a hand to work.

She and her friends were checking out the guys. For a couple of years, Lohill became a teeny-bopper's hot spot.

Good Times, Too

Like every brother and sister, there were times when Mom and Dad caught us wringing each other's necks. But we had a lot of good times neither one of us has ever forgotten. When Janet joined 4-H and wanted to show cattle, I taught her how to lead Blossom, her prize Holstein.

Janet later was a fifth grade teacher at the school she attended as a kid a few miles from the farm. And since I've left the farm, she's kept our tradition of having the same family running the farm for over 140 years.

In fact, excluding her college years at Michigan State University and the year or so she spent in Germany when her baseball playing husband, Neil, was in the service, Janet has spent nearly her entire life on the farm.

"I knew I wanted to stay on Lohill," she says, "but I assumed whomever I married wouldn't want to." As it turned out, she married a city boy from Pontiac who spent a lot of time on his uncle's farm and always wanted to farm himself.

When she was in college, she helped us out all the time babysitting our first daughter while my wife finished her student teaching. She and I held my Mom's hand together the day she died, and she's been looking after Dad for years now.

My mother-in-law always describes Janet as "the nicest person she ever met."

All jokes aside, Janet, I think you're OK. Here's looking at you, Sis.

Friday Nights, Sunday Mornings, Chocolate Sodas

WHEN I WAS A grade school youngster, we always kept a small flock of 100 or so laying hens at our family's Lohill Farm. Later, the flock grew to the point where we had as many as 4,000 laying hens at a time.

I fondly remember those early days when it was my job to feed and water the hens and gather eggs after school. And to later help Dad candle eggs, grade and place them in cartons.

Some customers came to the farm each week to pick up a dozen or two of eggs. We also had a half dozen customers to whom we delivered eggs in town on Friday night and another half dozen folks who got eggs delivered along with their Sunday morning newspapers.

As a youngster, I always tagged along when Dad went to town on Friday night to deliver eggs. Besides several homes, we delivered eggs to the two highly

DELIVERY TIME. Even with over 4,000 laying hens, Dad still continued the tradition of delivering several dozen eggs to a half dozen long-time customers who lived in Lake Orion.

GOOD TIMES REMEMBERED. While this was not the identical store, this soda fountain and bakery counter was similar to one where several generations of our farm family enjoyed stopping in a nearby town.

competitive drugstores located in downtown Lake Orion across Broadway street from each other—Grigg's on the west side and Van Wagoner's on the east side.

The real Friday night attraction for me was always sitting down at the counter in Van Wagoner's and ordering a 25-cent chocolate soda—always with vanilla ice cream.

Became News Hound

While waiting for the high school kids behind the soda fountain counter to prepare my chocolate soda, I'd head for the magazine rack and pick out one of the latest sports magazines to read—something like the monthly *Sport* which was really popular long before *Sports Illustrated* made its debut.

I'd sit at the counter poring over the latest sports news while fully enjoying my chocolate soda—a weekly treat that I seldom missed.

Well, it wasn't quite a once-a-week treat which brings me around to Sunday mornings.

Dad and Mom always had me at the Methodist Church for the 9:30 to 10:30 a.m. Sunday School session. It wasn't always easy getting the chores done and

"I could hardly wait until the Sunday School class was finished..."

rushing around to get to Lake Orion by 9:30 a.m., but somehow we always made it.

After dropping me off, Dad

GREAT CHOCOLATE SODAS. This soda counter was similar to the one which was part of the Van Wagoner Pharmacy in Lake Orion.

would deliver eggs to the half dozen regular Sunday morning customers around town before going down to my Grandma and Grandpa Tarpening's to visit until the church service started at 11 a.m.

As a youngster, Sunday School was enough spiritual learning for me and Mom and Dad didn't make me sit through another hour of church services. So after Sunday School, I'd walk two blocks up the street to Van Wagoner's Drug Store. Yep, you guessed it—for another one of their great chocolate sodas.

Week after week, Sunday after Sunday, I'd order a chocolate soda in mid-morning. I'm sure I was always the first customer to order one of these ice cream delights at such an early hour on Sunday morning.

The Rookies!

Every once in a great while, there would be a newcomer behind the soda fountain counter who didn't know how to make a chocolate soda. He'd tell me he didn't know how to make one

> *"Ray himself would whip up one of those 'Frank Lessiter' special sodas..."*

and then I'd tell them to go ask the owner and pharmacist Ray Van Wagoner.

Ray himself would come over and show the new employee how to whip up one of those "Frank Lessiter specials."

While that was going on, I'd head for the magazine rack to pick out something special to read. Friday nights were reserved for sports reading. But Sunday morning meant more serious reading and agriculture was often my favorite topic at this time in the new week.

These Sunday morning chocolate soda episodes served as my introduction to *Farm Quarterly*

BEST OF FRIENDS. The Byron Chapin home was always an important stop after Methodist church services. The excuse was to deliver a dozen of farm fresh eggs to them each Sunday, but it always meant spending 20 minutes or more talking between members of both families. And sometimes it even turned into enjoyable Sunday afternoon picnics on the Chapin lawn.

At left, Helen Chapin and Mike Lessiter pose for the camera during a visit.

magazine which was sold on the drugstore newsstand.

I used to wait for the latest copy of this large format magazine four times a year and would devour the mighty long, in-depth, page-after-page articles that were always their trademark. (Later, I remember how *Farm Quarterly* came out six or seven times a year, but still stuck to its four times a year name before eventually going out of business).

After reading and finishing another fantastic chocolate soda, I'd walk the half mile to my grandparents' house to wait for my parents to return from church.

In my later high school years, I eventually gave up both Sunday School and those Sunday morning chocolate sodas to accompany my family to church services.

We still delivered eggs to a few longtime customers on Sunday morning, but one stop always took place after church.

That was to the home of Byron and Helen Chapin, life-long friends of my parents. Our entire family almost always went in for coffee, pop, cookies, a game of ping pong and great conversations...and to deliver a dozen of our farm fresh eggs.

Times Change

Now, when I go back to Lake Orion, both of the old drugstores are gone and other retail businesses are located in these old buildings.

But for years as a grade school youngster while growing up on the home farm, this was my twice-a-week ritual—a chocolate soda on Friday night and another on Sunday morning.

Even today, enjoying still another chocolate soda can always be a highlight of my week.

RETRACING STEPS. Grandma and Grandpa Lessiter show grandson, Mike, a few of the stores in Lake Orion where the family did the majority of their trading for many years. And somehow the tour included the old Van Wagoner Drug Store and the stories about how Mike's Dad always enjoyed those great chocolate sodas on Friday nights and early Sunday mornings.

Oh No! A Snake In The House!

SNAKES HAVE NEVER been among my favorite animals.

In fact, I was plenty scared of them growing up as a kid back on the farm and that certainly hasn't changed over the years.

Whether it was a 7-foot long blue racer slithering between my legs as I brought the cows up the lane or a rattlesnake hissing through his teeth as I shocked wheat, seeing snakes has always sent a chill up my spine.

Even the smallest garter snakes that darted out from among the tomato vines in the garden while I was weeding could scare the daylights out of me. Or a brown snake crawling out of a bale of hay as I picked it up to toss on the wagon could make my heart skip a beat.

Yet a few special snake stories come to mind when I recall my days growing up on the farm.

Haymow Rattler

One time a rattlesnake somehow found his way into the haymow. We never saw it, but you could hear him rattling when we went up in the mow to toss down hay for the cows.

It was scary because I knew someone might end up with a nasty snakebite. As a result, nobody wanted to go up to the mow to toss down hay.

Except Dad, who never let things like this bother him.

Finally, he'd had enough of the hired man and a young teenager (me) being so worried about this rattlesnake living the life of leisure in the haymow.

So Dad and Harry Robertson, our hired man, grabbed pitchforks, hoes and a pair of coal stoker tongs and headed for the haymow to do battle with the lone rattler. When Dad told me to stay down below, I certainly didn't argue.

After 20 minutes of sticking the pitchforks here and there, they heard the snake rattling and Dad somehow grabbed him with the stoker tongs.

I still remember hearing this loud chuckle from Dad, then having him toss the dead rattlesnake right at my feet on the barn floor with a quick flip of those coal tongs. Even that scared the daylights out of me!

Rattle Grabber

There was also a time when I was riding my bike down Baldwin Road when I spotted a car parked by the side of the road. A man was standing near the road's center line.

As I rode closer, I noticed he had cut a branch from a nearby tree and was using the stick to hold down the head of a hissing rattler. With the other hand, he was pulling all those rattles off the tail.

Talk about scary! And dumb! If the stick slipped off the rattlesnake's head, he would have had a serious snakebite to worry about.

I rode away on my bike, shaking my head in utter amazement

THE CORNER OF THE HOUSE next to the back porch shown behind Mom was where the snake got into our house. Mom and Dad weren't smiling as much that day.

at what a crazy city slicker would be willing to try.

Snake In The House!

Mom also shared my fear of snakes—in fact, maybe that's where I got it. When I was 6 years old, a piece of siding had been off the house next to the back porch for several weeks.

I still remember what happened next very well. Mom came out of the house one bright summer morning and saw a snake sunning itself on the back porch. She screamed, then had me run and get Grandpa Frank to come quickly to kill what she felt was a rattlesnake.

After he got there, Grandpa started laughing as it was only a garter snake and not very big at that. He didn't think anyone should get too upset about a small snake and told Mom—absolutely the wrong thing to do.

Mom kept after him and he finally tried to smack the snake with the hoe. But Grandpa was laughing so hard the snake somehow got away and slithered off the porch. Yep, you guessed it—the snake crawled through the hole where the siding was missing and into the house.

Mom was scared to death. She feared she'd wake up in bed one night and find the snake under the covers or find it slithering across the kitchen floor while she was getting dinner ready. Plus, she was really mad at Grandpa Frank for letting it escape into the house.

By this time, she wanted Dad there—now! So I rode my bike down the lane to the field where he was mowing hay and told him he'd better get to the house pronto.

Not much ever made Mom mad. But she really let Dad have it for not putting the piece of siding back on the house weeks earlier.

He grabbed a hammer and some nails and started to put the siding back on, but Mom stopped him in his tracks. She told him in very few words that the hole might be the only way for the snake to get out of the house.

Dad didn't have the foggiest idea how to coax the snake out of the house. And nobody was even sure it was still in there.

Mom didn't go down to the basement alone for a couple of weeks to do the laundry.

The snake never did show its face again. After a month's time, Mom decided it was no longer in the house and things started to get back to normal.

Farming Can Be A Big Blast!

IT TAKES ALL KINDS of talents to run a farming operation, but dynamiting isn't one you'll often find every farmer up and down country roads putting to good use.

That's one reason I remember when Dad used dynamite to do some mighty hard work in a hurry—things that would have taken days and weeks to do otherwise.

Week Long Learning

When he was a student in the 1930s at Michigan Agricultural College (what is now Michigan State University), the agricultural engineering department offered a 1-week course in using dynamite on the farm. So he stayed in East Lansing for a week over Easter vacation during his junior year in order to attend dynamiting school while picking up a few more college credits.

What he learned was plenty of practical information which he was able to put to good use many times during his lengthy farming career.

Leave It Alone!

My first experience with dynamite took place when Dad showed me what was kept in the big red livestock show trunk that sat in the back of our farm shop. That's where he kept the sticks of dynamite, caps, electric wires and a 6-volt storage battery.

He showed me where these materials were kept and where he stored them for a good reason—so I'd stay away from them.

As promised, I didn't touch or play with it. But I'd occasionally lift the lid on the old trunk and take a look at the dynamite (and sometimes show my friends).

Actually, I probably had a much better idea of how many sticks of dynamite and caps Dad actually had on hand than he did, since I kept looking in there all the time.

Rock Work

I was 10 years old the first time I saw Dad work with dynamite. An enormous rock had worked its way to the surface in one of our hilly fields and the hired man had struck it while plowing. One way or another, it had to come out.

After spending a few minutes

"It took time, but Dad knew what he was doing and he wanted it done right..."

---

trying to dig it out, Harry and Dad realized the rock was practically the size of one of today's big round hay bales. The decision was made to dynamite it.

I still remember helping dig under the rock in several places so half sticks of dynamite could be properly placed. It took time, but Dad knew what he was doing. And he certainly wanted it done right.

Dad let me uncoil the 250-feet of electrical wire as we walked away from the rock where the dynamite had been placed. The two of us laid down on the plowed ground and kept our heads close to the ground while he attached the two wires to the 6-volt storage battery.

It wasn't like in the old Western movies where a cowboy simply took a match and lit a piece of fuse, then ran for safety before the whole town came tumbling down. And we didn't have a modern-day plunger box that showed up in many of the later movies either.

Wham! Bang!

Dad simply touched the bare ends of the two copper wires to the battery terminal. After the tremendous noise from the explosion, pieces of rock were flying everywhere. Even 250 feet away, small fragments of rock rained down on our heads.

When we went back to look at what was left of the rock and the newly dug hole, there wasn't much. There were a few large pieces of rock laying around the field which we later hauled to the fence lines.

But basically, the dynamite had shattered and destroyed the big rock that had been in this field since the Ice Age.

I was certainly impressed with my Dad's dynamiting abilities. He'd saved plenty of time and hard work in eliminating that rock with very little effort.

It's My Turn

Just a couple of years later, another big rock sprung to the surface in a nearby field. Again, it was much the same story, with one important change that meant a great deal to me as a farm youngster.

This time, Dad let me touch the wires to the 6-volt storage

"The loading dock had to make way for the milkhouse, but it certainly wasn't going to be easy to remove..."

---

battery. I was the one who proudly got to wreak havoc on that rock.

Loading Dock Woes

Dynamite also came into play a few years later when we wanted to build a new milkhouse next to our big barn. The older, much smaller milkhouse where cans of milk were cooled in large open tanks was located 20-yards uphill from the barn.

With more cows in the milking string, we had outgrown the old building and the decision was made to build a larger, more modern, more efficient milkhouse much closer to the big dairy barn.

But there was a problem. A solid concrete loading ramp had been built decades earlier into the side of the hill right smack where we wanted to locate the milkhouse. It stood 40-inches tall and measured 8-foot square. The big question was going to be how to remove it.

Dad decided the job could be done with dynamite, even though the loading ramp stood less than 10 feet from the corner of the big wooden, stone banked barn. Plus, there was also a corn-crib and storage barn less than 30 feet across the driveway and

LOADING DOCK ROCKS. Located right next to the big dairy barn, the truck loading dock consisted of plenty of rocks and concrete. Yet it came apart fast when dynamited and without any damage to the barn.

other nearby barns and silos.

When he talked about dynamiting the loading ramp, I was scared. In fact, I really doubted whether it could be done without damaging the barn.

But Dad was sure it could be done—providing the right amount of dynamite was used, it was properly placed and we didn't try to do it all in one blast.

While the rocks we destroyed earlier in the fields always went

> *"We used small pieces of a single stick at a time to control the damage..."*

up in one big explosion, this was a much different matter. Dad wanted to take it slow and make sure there wasn't any damage to the barns.

I don't remember the exact number of dynamite sticks that were used or how many blasts were made, but Dad kept it to a minimum—using small portions of a single stick at a time to dismantle the loading dock in several steps rather than in a single shot.

I recall stringing wire around two other buildings and then having small pieces of rock rain down on our heads when the dynamite went off.

But it did the job. Before long, the old loading dock was a shambles of small concrete chunks which could be readily carted off. It wasn't long before a new milkhouse was standing where we used to load livestock for hauling to market at the Detroit Stockyards or to be taken to the county and state fairs.

Dynamiting A Ditch?

One other dynamite experience I always heard about on the farm happened when I was much younger. It showed what dynamite could really do and also showed Dad's ingenuity when working with it.

He and Grandpa Frank decided to drain a 10-acre muck field located at the end of the farm lane. This permanent pasture had been used for years for grazing beef, sheep and dairy cattle, but they wanted to grow some high-value row crops there—like raising early-season potatoes.

MILKHOUSE REPLACED DOCK. Built directly into the side of the hill next to the dairy barn is the milkhouse. The previous truck loading dock was dynamited into pieces and hauled away. It was a real scary time, but Dad was convinced proper dynamite placement would avoid any barn damage. As usual, he was right.

DYNAMITED DITCH. Shown some 30 years later, this is part of the 400-yard long shallow drainage ditch which was used to drain 10 acres of the farm's muck ground. Just a touch of the wires to the battery and the dirt was flying everywhere, thanks to proper placement of the dynamite sticks.

At one end of the field, a tile drainage system had been installed years earlier. This drained into the lake located about a quarter mile away.

Somehow Dad convinced my Grandfather Frank that they could use dynamite to carve out a drainage ditch along the western side of the field. I don't think Grandpa Frank was fully convinced it would work, but he let Dad give it a try.

Using what he'd learned in school and carefully placing half and quarter sticks of dynamite along the entire field length, Dad blasted a 400-yard long shallow drainage ditch with one quick

"I would have really loved to see this ditch being formed out of the soil as all of the dynamite charges went off..."

touch of the two wires to the storage battery. It was done without any costly shoveling or hiring a drainage contractor to dig out the ditch.

As farm folks told the story over the years, the ditch was soon draining into the existing tile system and the high water level in the field went away.

Potatoes and other high value crops were soon seen growing profitably in this extremely valuable, high organic muck soil.

I'd have loved to have seen this take place. Many times over the years, I managed to leap this drainage ditch and marveled at the fact that dynamite helped shape it without anyone even lifting a shovel.

Summing Up

Dynamiting certainly isn't for every farmer and you definitely need to know what you are doing. And it certainly isn't something kids should be playing around with either. Today, it's probably best left to blasting specialists.

But I was certainly impressed as a kid with the talents Dad showed in letting a few sticks of dry powder take over for plenty of long days of back-breaking, back-aching labor.

Trees Really Made The Farm

IF YOU ARE used to tilling fields where you can see for miles, don't want to worry about going up or down hills or are concerned about working farm machinery around trees, then

GORGEOUS FALL COLORS. Generations of family members say the hundreds of trees found on the farm are especially beautiful during the late fall harvest season.

living at Lohill Farm certainly wouldn't be for you.

The front two-thirds of the farm could be considered slightly rolling without too many farming problems. But several fields toward the back of the farm contained some of the steepest hills you'll find anywhere.

Most of these slopes were farmed, but a few were simply too steep and were left alone for safety's sake. And when I found myself scared stiff as a kid riding the tractor up or down a few of these hills, I often wondered why Dad didn't leave a few more of these steep hills in permanent grass.

Yet the rolling fields made a pretty site when the land greened up in the spring, the crops started growing and the small grain and corn crops turned from green to brown to gold in late summer and fall.

TREES EVERYWHERE. Wooded areas and field borders on the farm are filled with many different tree varieties.

GROWING OLD. Like the six generations of family that called Lohill Farm home, trees growing on the farm also went through a similar life cycle.

Thousands Of Trees

Almost all of the farm's borders are still wooded with thousands of trees. Much of the lumber used in the farm's buildings came from trees cut by fam-

ily members from several of our woodlots over the years.

Several times, a portable sawmill was moved into the woods and freshly cut lumber was hauled away for stacking. At other times, horses hauled logs out of the woods to nearby sawmills.

Once or twice, logs from valuable species of trees found in the woods were sold to forestry companies after being marked by state foresters.

Scrap wood from cut timber and from dead trees found around the farmstead was used as a low-cost way to heat the original farmhouse for more than 100 years. A winter ritual was using a tractor-powered buzz saw to cut logs and dead wood into lengths that could eas-

FENCE ROW TREES. When the land was cleared more than 125 years ago by family members, many trees were left along fence lines. Later, many trees which sprouted as saplings in the fence row areas were left to grow in order to provide valuable shade for the farm's livestock.

ily be tossed in the furnace.

The Beauty Of Trees

It's still a real delight to check out the different species and ages of trees in the farm's woods.

The same is true of trees growing along the farm lane and the few remaining maples left along Baldwin Road that were transplanted as young saplings from the farm's woods by Grandpa Frank and his brother, Floyd, more than 80 years ago.

Many generations of adults and youngsters alike have enjoyed the true beauty of trees to be found at Lohill Farm.

MANY FARM USES. Over the past 125-plus years, timber was harvested several times from the farm's woods to provide lumber for building barns. Waste wood was used for heat and an occassional timber sale was made to the area's sawmills and timber brokers.

Real Beauty In Farm's Old Barns

THE SUPERIOR craftsmanship that went into building older barns is now a thing of real—yet quickly vanishing—beauty.

As a result, nobody builds barns like they used to.

And I'm sure many farm builders would say, "Thank goodness for that."

Modern day building crews simply aren't geared to tossing an axe over one shoulder and marching off to the woods to start carving out another barn from growing trees. Even the environmentalists would get alarmed today when they hear talk about all this timber cutting.

A Great Old Look

Yet there's a certain sense of beauty, craftmanship, individual design and unequaled handwork that went into many of the older barns that still dot our countryside. Unfortunately, many of these old barns are becoming a vanishing landmark as more and more disappear from the rural landscape each year.

Why did so many farmers like my ancestors want such big barns in those old days? They needed plenty of storage. They needed a way to handle loose hay with forks which were controlled by ropes that ran to a rail at the peak of the barn's roof.

Grain bundles were also hauled into these barns during July and August to protect the valuable grain from the weather

BARN BUILT IN 1905. Plenty of craftmanship went into the construction of this barn, long a mainstay on the family farm.

THREE YEARS LATER. Except for going from red to white, the big barn didn't change much in more than 90 years of usage. Shown above in 1908 with one of the farm's prize Shorthorn bulls is Frank Lessiter decked out in his Sunday best church clothes.

while waiting for the threshing crew to arrive.

90 Plus Years...Still Strong

Anytime I walk into the big barn (96-feet long by 40-feet wide and 60-feet tall at the peak) on our home farm in Michigan, I have to marvel at all of the beautiful craftmanship that went into construction of the barn.

Erected in 1905, the barn has been a real mainstay to the success of the Centennial Farm and has made many unmeasured contributions to the success of our family's farming operation over the past 90 plus years.

When we make the long trip back to the Michigan farm, I like to take our four children and grandchildren out to the barn to show them the fine craftmanship and history that goes with our barn. With a sharp eye, you can see where the lower timbers take on a special honey color and smoothness where they were softened by the rubbing of hay and straw brought in on wagons

LOTS OF BARN DETAIL. The early-day carpenters always insisted on adding plenty of details when it came to building these beautiful two-story barns.

HAYMOW CRAFTMANSHIP. All of the 8-inch square cross beams used in the barn were hand cut from logs which were at least 60-feet long. The beams, crossties and supports were held together with round pegs rather than nails. Over 90 plus years, no peg ever failed to hold the beams.

PLENTY OF WEAR. Nobody knows how many feet were placed on the rough, hand-built barn ladders over the years.

Many Red Barns

Like many early day barns, the big banked barn at Lohill farm started out in life painted red (see page 84).

It wasn't that farmers all over the country loved red. Instead, commercial paints were very expensive in those days and farmers needed something they could put together in large amounts for use on their own barns.

The standard red paint recipe in olden days went something like this:

★ 1-gallon of skim milk.

★ 12-ounces of lime.

★ 8-ounces of linseed oil or neatsfood (also known as cow's hoof glue).

★ 3-pounds of red oxide of iron.

★ 4-ounces of slacked lime, oil and turpentine.

BANKED BARN STYLING. After the big barn was completed in 1905, an earthern bank was built on the west side to allow horses to pull wagons loaded with hay, small grains and other feeds into the upper area for winter storage.

over 90 years of use.

Then there are the beautifully fitted beams, morticed into posts and held with wooden pegs. And if you look hard, you can still see the scribing lines a carpenter made on the beams in 1905 when he was trying to fit them together.

90 Plus Years Of Storage

As you notice the steel track in the peak of the barn where ropes for the hay forks were attached, there's no way you can imagine how many thousands of tons of loose or baled hay were stored in the barn over the years. Or the

WEATHERED MANY STORMS. The old barn and other buildings on the farm always came through the winter with flying colors, thanks to excellent construction.

OLD SHEEP BARN. This long-gone structure was actually an extension to the main barn. It was used for many years for housing the farm's flock of sheep which was part of the diversification efforts in the farming operation.

WORN OUT, TORN DOWN. The old three-section sheep barn finally outlived its usefulness and was torn down in the late 1960s. Some of the barn's old beams eventually found their way into construction of expensive houses while some barn boards ended up in the home of one of the Lessiter family members in northern Illinois (page 99).

TORONADO DAMAGE. Lumber was scattered everywhere after a toronado hit the old wooden silo located at the north end of the barn. A used redwood silo was soon purchased as its replacement. Below and at right is a concrete stave silo which stood at the barn's south end.

number of bushels of wheat, barley, rye and oats that were threshed on the two dry floors of the barn and stored in the second floor granary built under one of the hay lofts.

Up until the early 1900s, the family made do with several smaller two-story timber frame sheds with dirt floors. But all of the hay and straw had to be pitched through small second floor doors—which was long, hard work for the farm workers over several decades.

But with the increasing size of the farm's Shorthorn herd, a bigger barn was definitely needed.

After the decision to build the

POTATO STORAGE DOWN BELOW, MACHINERY STORAGE UP TOP. Built into the ground, this two-story barn provided plenty of storage space and good insulation for the winter processing of the farm's potato crop.

big barn was made in 1903, the family spent 2 years cutting oak trees from the woods on the farm. This timber was destined for beans, haymow flooring, framing, rafters and roof boards in the new barn.

Snaking Trees

You have to marvel at the work that must have gone into cutting those big trees and "snaking" them with horses to the sawmill set up on the farm.

Those were big trees...they had to be with 14 beams in the barn measuring 40-feet long and 8-inches square. After cutting, the green beams had to "season" for at least a year before actual construction could begin.

Several acres of timber, located a mile north of the 175-

THE OLD FARM SHOP. For many years, hand and power tools, equipment and supplies needed to keep everything operating smoothly were kept here. It also included a fire pit and huge pot where water could be heated for butchering hogs.

CLOSE TOGETHER. Most of the farm's barns were located fairly close together. Especially in the early 1900s, this made it very convenient to move feed, animals and equipment.

acre family farm were purchased to cut wooden siding for the big barn. With a permanent sawmill situated on a nearby pond, these trees didn't have to be "snaked" far for cutting.

Besides timber, tremendous amounts of stone were hauled for the west wall of the barn. Built into a man-made hillside, this wall measures 96-feet long, 8-feet high and 18-inches thick. Other stone was hauled for the foundation and basement floor.

Finding stone was no problem with the many rocky fields that dotted our family farm.

Like many other barns of its time, the barn was built into the side of a short, steep man-made hill. This popular building concept enabled farmers to drive a team of horses and wagon up the hill right onto the second story floor of the barn into two of the

> *"Carpenters showed up on Monday morning and worked through until Saturday night..."*

16-foot wide bays which featured big sliding doors. Livestock were housed on the lower floor which was built into the side of the man-made hill.

Once actual construction of the barn began, a team of carpenters from a neighboring town arrived each Monday morning. They would work until Saturday night before heading home for a short weekend.

Framing for the barn was laid out on the ground, cut and fit so each of the many pieces properly went together. It's likely neighbors from miles around must have helped in the actual raising of the barn.

It fell on the shoulders of my grandmother to feed the carpenter crew and to provide them with beds all during the week.

200 Families Celebrated

Completion of a barn in the old days was nearly always a cause for a big celebration. My grandparents happily recalled many times how more than 200 couples showed up for the big barn shindig soon after it was completed.

There were horses and buggies tied everywhere at the homeplace that Saturday night. Even the children came...and they were bedded down for the

> *"Stairs were built up to the hay mow over the grainary so the kids had a place to sleep..."*

long night in the haymow above the granary.

While the haymow over the granary is 10-feet higher than the main floor of the barn, stairsteps were built up to the mow just for this "barn warming." Just like they did back home, the children simply walked up the stairs to bed.

Two orchestras played for dancing in the barn all through the night.

Plenty of good things to eat were laid out on a 4-foot wide by 25-feet long buffet table. This table was to eventually see better than 60 years of use.

Built originally by the barn carpenters, the table was used to lay out small framing materials during construction. In later years, it doubled as a summer party table and as a handy snow fence during the winter months.

Tough Old Silo

A 16-foot diameter by 36-foot high wooden silo was built at the north end of the barn in 1910. One of the first silos in the area, it took plenty of corn silage to fill. In fact, my Dad remembers many times when it seemed like the green corn silage would

A BIG, BIG BARN. Measuring 96-feet long, 40-feet wide and 60-feet tall at the peak, this two-story barn was one of the bigger barns ever built in the Lake Orion area.

practically settle as fast as the silo could be filled.

There weren't any silo unloaders in those days—just my Grandfather, Dad and the hired man. It was real hard work on a cold wintery morning to go up there and pitch enough corn silage with a fork across the silo and still hit the silo chute.

The barn withstood many years of hard use. In the late 1920s, the barn came out on top in a battle with a fierce toronado. However, the wooden silo built at the north end of the big barn was destroyed.

"We couldn't see the silo from the big house," recalls Dad. "So we didn't know it was down until somebody driving by stopped to look at the damage and then came to the house to tell us. It happened during the summer, so the silo was empty.

"I remember going out there and seeing the silo sprawled all over the field. It was a real mess. There was lumber everywhere and silo hoops were bent in every direction. Many of the white pine silo staves were smashed to pieces."

Eventually, a used redwood silo measuring 14-feet in diameter and 30-feet high was purchased to replace the destroyed silo. Some white pine from the original silo destroyed by the tornado was salvaged, so the new silo was actually built from a combination of white pine and redwood. Later, a concrete stave silo was added at the south end of the barn.

The wooden silo was still going strong in the early 1960s when it was pulled over with a tractor since we didn't use the silo any longer for silage. However, we couldn't sell the old silo and had another use for the lumber. Dad wrapped a rope around

AIR DRIED CORN. For many years, the farm's corn cribs were used to dry both hand and machine picked ears of corn.

the silo about halfway up, hooked it to the tractor and slowly drove away. The silo toppled over with very little damage to the wood.

The tongue and groove lumber from the silo was used to resurface the two second story "dry floors" in the big barn where we drove up the hill with tractors and trucks.

The barn was also reshingled in 1935 and the roof took 65 squares of shingles to cover.

The original shingles had not worn out, but the nails holding the shingles had rusted out. The three big ventilators on top of the barn were also installed when the new shingles were installed.

Those shingles have already lasted more than 60 years and the barn roof still doesn't leak.

Special Beauty

As beautiful as these old barns are, they are fast disappearing from the countryside. There's a special beauty in these old barns...a symbolic sense of security and warmth.

Livestock were secure and warm when housed in these well-built barns even during the coldest weather. And even to this day, a wooden barn can be a comforting place to be on a miserably cold and blusterly midwinter day.

So the next time you come across one of these old barns, take a few minutes to walk inside. Admire some of the beautiful craftmanship that went into these barns that still rank among some of the finest buildings to be found anywhere in the world.

Barn Wood Has Special Meaning

WEATHERED SIDING from one of the old Lessiter barns ended up on the basement walls you see here. That was after making a 308-mile journey to Mount Prospect, Illinois.

Need Something Special

When it came time to panel a basement recreation room, one

"More than 60 years of wear gave the barn siding a rugged rural look..."

of the Lessiter kids living in a northwestern suburb of Chicago decided it would be great to use barn siding. The family wanted something unique—something with a touch of family history that would make this basement recreation room very special.

The old 24 by 60-foot sheep barn on the Michigan family farm had not been painted in more than 60 years and the weathered siding had that rugged rural look to it. It had been

SHEEP BARN TO REC ROOM. The weathered siding installed on the walls of this basement recreation room made a 308-mile journey from the Michigan family farm to this home in the surburbs northwest of Chicago, Illinois.

NEAT PLAYROOM. Besides providing plenty of entertainment for the entire family, the unique weathered look of the basement recreation room always brought back many fond memories of the home farm in Michigan where the siding had once covered one of the old barns.

cut from trees growing in the farm's own woods more than 60 years earlier.

Made Mom Happy

The old weathered barn had hardly been used over the past 10 years and was suffering from neglect. Mom had even declared it as a "public eyesore" and had

"Mom was delighted to see us haul away the siding..."

been after Dad for several years to tear it down.

So we went back to the farm one spring weekend and dismantled the old barn. As we pulled the 17-foot long boards off the barn, we precut the siding to a 7-foot length so it would later fit perfectly when nailed on the basement walls in Illinois. We borrowed one of the farm trucks and hauled the meaningful wood siding to its new home.

The end result was being able to enjoy the family's own personal old weathered barn in this basement.

Machinery Upstairs, Potatoes Downstairs

LOOKING AT THE OLD truck garage, you'd think it was just another one-story building used to store farm machinery during the winter.

But the old building was used for much more than that.

Even so, many people didn't realize the building also contained a dark, humid cellar with plenty of high ceilings and immense storage space.

Built into the side of a hill, the building had been designed for machinery storage on the upper floor along with a potato and vegetable cellar. Yet we never raised enough acres of potatoes to fill the cellar in the winter.

When I was growing up, the people who ran the 1,200-acre

NO ORDINARY BARN. The truck garage served many unusual purposes which folks didn't realize passing by on the road.

nearby Lakefield Farms always raised 40 acres of potatoes across the road from our farm.

Because this building was close for storing potatoes coming out of the field, and since they didn't have enough potato storage of their own, they leased this cellar each year.

By the end of October, the cellar would be jammed from top to bottom with 40 acres of spuds.

A wood-burning furnace and brick chimney provided heat to keep the potatoes from freezing. We'd also bank the outside walls of the barn in late fall with manure which provided insulation.

Cold Weather Work

Once the weather turned cold, the Lakefield Farms crew would start working the potato crop.

By mid-morning, you'd see three or four pickup trucks parked in front of the old truck garage. Since drivers on the highway would see the trucks but nobody around, they wondered what was going on here.

The workers parked near the doors on the upper floor since there was a trap door in the northwest corner of the building. It opened onto a steep set of stairs which led down to the potato storage area.

As a kid, I used to enjoy going down into this warm area where the men would be cleaning, grading and bagging potatoes in 50-pound bags for the Detroit grocery store market.

The only time the big insulated doors on the bottom level of the building were opened was when a semi-trailer would carefully make its way through the barnyard. Backing into the potato cellar entrance, the men would load the truck with spuds.

Great Looking Chicks!

When Lakefield Farms was sold in 1955, it was the end of storing potatoes here. For a number of years, the cellar area was used for storage of machinery.

But as Dad moved into the egg business on a big scale, the cellar area was needed to start out as many as 1,000 newborn chicks.

To keep these young chicks warm, we relied on an oil-burning furnace to keep the cellar area especially warm. We also used heat lamps to keep the chicks from becoming chilled.

Using this cellar for rearing these birds meant lugging heavy pails of water from the milkhouse which was 75 yards away. And making several daily trips hauling both heavy bags of feed and pails of water down that steep flight of stairs wasn't easy.

At first, the chicks had to be fed and watered every few hours around the clock. While Dad handled the feeding and watering during the day, it was my job to make the final check around 10 p.m. each night. I had to make sure the chicks had plenty of feed and water and that the temperature was high enough to keep a chill off the birds until morning.

Not Just Machinery

So there's much more to the history of that old truck garage than just serving as a place to store machinery.

The Upside Down Car

One cold, blustery winter night, I was shivering in the milkhouse about 10 p.m. while filling a bucket with water. The water was for the 1,000-plus young chicks we had just started out as future laying hens in the old potato cellar.

Out of the corner of my eye, I saw a light which seemed to rotate in a circle flashing through the window. But with my mind on the upcoming Friday night basketball game, I really didn't pay much attention. Maybe it was just my imagination playing tricks on me!

Once the buckets were filled with water, I made my way to the truck garage and fed and watered the chicks for the last time on this night.

So I was surprised 15 minutes later when I walked around the corner of the farm shop and saw the flashing lights of police cars and fire trucks bouncing off the road in front of the barn.

As it turned out, the circular motion of the lights I had seen earlier had been real. A driver had been heading down the icy road when his car slid sideways. The auto flipped in the air and did a complete circle out on the road—the lights I saw flash through the milkhouse window.

Nobody was seriously hurt, but I was shaken up when I learned what had really happened.

Everything had been so quiet when I'd gone outside to feed and water the young chicks that I'd never noticed what was going on out on Baldwin Road.

Worrying About The "Barn Burners"

UNFORTUNATELY, IT happened several times while I was growing up at Lohill Farm.

The word would quickly spread through the Lake Orion, Mich., area that some of the area's big barns were mysteriously going up in flames.

Several times, it happened during the haying season, which led a few non-farm folks to believe farmers were storing wet hay which led to combustion and the resulting barn fires.

But that never was the case.

Instead, each time it was one or more deranged individuals who seemed to get a kick out of setting barns on fire...and then watching them burn.

As a result, Dad and other area farmers were deeply concerned whenever the word got out that an arsonist was on the prowl.

Haying Time

The time I remember best was one June when I was about 12 years old. Several outlying livestock and storage barns no longer in use on the Scripps farm, located only a few miles south of our farm had caught fire. Even though the Scripps had quit farming a few years earlier, it was still a great loss. And certainly scary to other farmers.

THE FEAR OF FIRE. Nothing was as scary to Dad as the news that "barn burners" were on the prowl.

While the Lake Orion fire chief suspected arson, he didn't give out much information concerning the fire.

A Barn A Week

A barn would catch on fire one week, then another barn would be found burning the next week. A couple of weeks later, another barn located somewhere in the area went up in flames.

With our own big barn full of hay in mid-June, Dad was extremely concerned about this situation. Anytime after dark when a car pulled into the barnyard, he'd quickly be up and out of his chair to peek out the window to see who it was.

Normally, it was somebody just turning around in the driveway. If they didn't get moving quickly, Dad would soon be heading for the barn to see what was going on.

Even though he certainly worried about the arsonist and our barns, it wasn't likely the "barn burner" would drive in our driveway, walk up to the barn and set it on fire. Yet you never knew for sure.

The arsonist set four or five barns on fire over a 2-month period before finally being caught.

As fire officials say, the best place to spot one of these guys is in the crowd of people standing

"The best way to find barn burners is to check out the crowds of people who are watching the barns burn..."

around watching the fire burn.

The arsonists often torch the barn, drive away, wait until the flames are shooting high into the air and later come back to join the crowd watching the barn burn. As a result, you often see police and fire squads photographing people standing around watching a building burn.

In fact, that's how they finally got a tip that led to the arrest of this guy—he kept showing up in the crowds of people who would gather to see the barns burn to the ground. After a serious interrogation at the county jail, he confessed to the fires. And all the farmers in the area went back to being able to relax again.

It is not a pretty sight to see a beautiful old barn end up in a heap of ashes and rubble. Built between the early 1900s and mid-1920s, these beautiful old barns were never replaced after they burned to the ground. As a result, an important part of America's past went up in flames.

Room Of Flames

Still another barn arsonist was making the rounds one winter in the southern Michigan community where I grew up when I was away at college. Dad recalls waking up in the middle of the night and seeing what appeared to be the walls of the bedroom in flames.

Talk about scary!

Instead, it was the reflection of flames from a neighbor's barn located 300 yards down the road which was burning—not our house. Dad called the fire department, quickly dressed, told Mom not to worry and headed for the fire scene.

Luckily, there were no animals in the three barns which ended up a total loss.

As it later turned out, the "barn burner" had been in the crowd as this barn burned to the ground.

WE WERE LUCKY. Dad says none of the Lohill Farm barns were ever lost to fire, although a wooden silo was once toppled over by high tornado winds.

HAY WAS A BIG CROP. Dad says fire was a major concern since many of the barns were filled with hay during the extremely dry days of summer.

When another barn went up in flames a few weeks later and this guy was present, the police nailed him.

Lightning Starts Fire

One other time, I watched as a neighbor's hay-filled barn went up in flames during a severe lightning storm. We lived a half mile away, but you could see the flames stretching 40-feet into the sky after lightning hit the barn.

Firemen saved a nearby old two-story banked barn that was used to store hay, straw and machinery on the upper level and farrowing sows on the bottom

"The heat was almost unbearable even as I stood well back from the burning barn..."

level. But the other barn used as a loafing shed for heifers and a nearby beautiful old-style glazed tile silo were destroyed.

I remember how hot that blaze was and how the heat was almost unbearable even as I stood well back from the burning barn. The tremendous heat popped those glazed silo tiles on the silo. You'd hear a loud bang as the air in the tiles was heated to the bursting point.

The Lonely Brother

Just a few years back, another barn burner hit our area. Michigan highway M-24 runs through Lake Orion south toward Pontiac. One afternoon, a friend of our family reported his barn was burning. By the time the fire department got to the scene, another barn was burning a half mile down the road.

By then, the police had wised up and they sped down the road to a third big barn located on the same side of the highway. Rushing into this third barn, they spotted a teenager pouring gasoline on bales of hay—about to torch his third barn in 30 minutes.

They say most barn burning arsonists are looking for attention, enjoy seeing buildings burn and often want to get caught. This was certainly the case with this teenager.

As reported later in the newspaper, an older brother was serving time in the county jail. His younger brother felt lonely and wanted to see him, but jail officials wouldn't let him in during visiting hours because of his young age.

He figured the easiest way to get in to see his brother was to set several barns on fire, get caught by the police and be placed in the same jail cell with his brother. That's exactly the way it happened, but it's sad that two old and beautiful barns had to pay the price just so he could visit his brother.

Fear Of Fire

Looking back, there were numerous barn burning concerns in our area of southeastern Michigan. And I can still see the worried look on Dad's face when these "barn burners" were running loose.

Dad always was careful about fires on the farm...never putting up hay that was too wet, making sure nobody ever smoked in the barns and not burning trash on windy days.

But there's not much you could do to protect yourself when a "barn burner" was on the loose...as they set fire to some of America's beautiful old barns.

There's No Place Like Home

EVERYONE ALWAYS SAYS you can't go home again. I disagree.

I drove out to interview a farmer in Michigan last week and decided to swing by and say hello to my sister and Dad.

As I drove along the winding Baldwin Road, I thought to myself, "this place has really changed since I was a kid."

There are now supermarkets,

THE OLDEST FARM HOUSE. Built in 1854, this was the original house on the Lessiter family farm. Remodeled several times over the years, it's still going strong and served for a number of years as the home for the Neil and Janet Roberts family.

THE BIG FARM HOUSE. Built by Great Grandfather John Lessiter in 1885, the house included six giant bedrooms, even though his family was practically fully grown at the time. The ceilings were 9-feet high and most of the bedrooms also had walk-in closets which measured roughly 6 by 8-feet. The outside of the big house also included four porches and many ornate carved decorations which required hours and hours of the carpentery crew's time to build.

A Four Porch House...

Most farm folks lived in houses with two porches—one in front and another in back. But the big Lessiter farm house had four porches!

The elevated front porch is what the family showed the world and it often suffered from a clutter of people. It was screened in during the summer and had storm windows installed during the winter months. Besides lots of comfortable chairs, it had a glider and a hammock which was a delight for a young boy like me.

This porch was used for entertaining guests and for sitting out on Sunday afternoons watching the traffic go by or for cooling off in the evening after a long hard day's work.

With top to bottom screens, it was always a summer delight and the perfect place from which to see all the traffic going by on Baldwin Road.

The extra long north side porch took on many of the duties of a typical back porch found on most farm homes. It was where the farm crew took off their muddy shoes before going in to eat.

It was the place to stay dry when it was raining cats and dogs only a few feet away. It was the place where kitchen deliveries of groceries were made. Unlike most porches, it had two doors that opened into the formal dining room and also into the kitchen.

The south side porch opened up from the back side of the dining room. It was a fine place to enjoy the morning sunlight. I remember when tasty grapes grew on a trellis along one side.

The back porch was nothing fancy and was hardly used. But it was located conveniently to go out and bring in wood for the cookstove, ice for the icebox or to take something down to the huge basement under the big house.

FEW CHANGES. In the 60 years leading up to 1955 when the above photo was taken, few changes were made to the big farm house. At left, Janet poses with her Holstein heifer calf, Blossom, near the house's unique front porch.

fancy subdivisions, championship golf courses—even a ski resort in the area—where before there was nothing but farmland.

But after setting the old station wagon into park on Lohill's coarse gravel driveway, I realized that very little had really changed since I was a kid. Sure, the windmill and some of the old barns are torn down, my sister renovated the tenant house and a school teacher now rents the house where I grew up.

But at a glance, that's about it.

In fact, except for more traffic on Baldwin Road and the fact that another house now stands where my grandparents' big house once was, people tell me things have pretty much been the

GREAT VIEWS FROM HOUSE. Snow-covered pines growing along the north fence of the yard provided beautiful winter views. The house sat across Baldwin Road from the barns, always providing a delightful look at what the men were doing in the fields and around the barns.

same there since the turn of the century. Except for the traffic!

The Big House

My Great Grandfather John Lessiter built the big house in 1885. Although his family was practically all grown up, it had six giant bedrooms. The ceilings were 9-feet high and five of the six bedrooms had walk-in closets measuring 6- by 8-feet.

The house was ahead of its time—having central heat from the very beginning from burning 4-foot logs. It didn't work too well, so in 1908 the system was replaced with a cast iron boiler. Pipes were run throughout the house and radiators were installed in every room—it was

GONE IN A FLASH! Only a few years after the big house had been sold, it was gone in a July 13, 1965, fire which also destroyed the nearby outbuildings that had contained the ice house, summer kitchen and garage in earlier years.

1952.

July 22 1965

A Landmark Is Gone

(Review photo)

One of Orion Township's landmarks was destroyed by fire on July 13. It was the home built by John Lessiter (grandfather of Orion Township Supervisor) at 501 Baldwin, then the Stream Mill Road.

White Pine from a woodlot on Stanton Road, part of the area known throughout the nation as the "Pinery" because of its density and size, was used in its construction in 1885.

The Lessiters purchased the property in 1855 and it remained in the family until April of 1964 when it was purchased by Mr. and Mrs. William Cole.

The fire started in the rear section of the building, originally built as an ice house for storage of ice from the lake at the rear of the property. It spread to the "summer kitchen" area of the huge house, and was out of control by the time the first fire unit from Gingellville arrived at 5:15 p.m. Lake Orion, Oxford Independence and Brandon departments also aided in the fight hampered by high winds.

So much water was drawn from the Judah Lake wells that residents were without steady water supply until 10:30 p.m.

Danger of the fire spreading to a grain field adjacent to the building to the North and the John Lessiter farm across the road made it necessary to keep everything well soaked.

The Coles have three children under six years old, who fled with only the clothes on their backs but were uninjured.

Cole was able to save both their cars, however, and they are living with a sister, Mrs. Mary Day, on Dixie Highway. They plan to rebuild on the site.

All the remains of the workmanship of the fine home are the hand hewn rock foundations — and a garden with beans ready to pick and one the best tomato patches in the area which the spectators carefully walked thru.

Perhaps the children will enjoy their swings that didn't even have the paint blistered the more, because of not having their other toys.

Many people have been generous, and the Lake Orion's Club have donated $100 to help the family.

The home and its contents were valued at $20,000, which was partially covered by insurance.

SAFE AND SOUND. For many years, this big safe sat in the front office at the big house. It was later moved to other houses on the farm. The heavy safe was moved on wooden planks laid over the stairs and pulled with the cable on a tow truck. John Lessiter shows grandkids Trista Roberts and Mike Lessiter how the old safe works and gives them an exciting look at various family valuables stored inside.

considered a good heating system for its day.

In later years, they installed a stoker and burned coal. Later when the boiler cracked and started to leak, they took the 1,400 pound furnace up the back stairs in one piece and installed a new Kewanee steel boiler.

I remember forking cow manure into 3-foot tall piles against the house during the winter for much needed insulation in the drafty old house—a common practice which Mom never allowed at our house.

Guests Always Welcome

Overnight guests and boarders were common at the big house. Bed space was never any problem, since all six bedrooms had double beds and there were couches and davenports in every room. Dad says sometimes there'd be more than 20 people staying at one time and those traveling by horse and buggy often stayed 2 or 3 days.

Since there were often last minute potential cattle buyers who stayed overnight, Grandma never really knew how many folks there would be to feed and bed down.

Dad also remembers the wanderers, or "tramps" as they were called in town, who traveled with old hats and carried a bundle under their arms—often their only belongings.

Many were poorly clad, with clothes patched in many instances and usually wore old hats. Most had a couple of day's growth of beard, but some were slicked up as best they could, says Dad. They were always in need of a meal and possibly a place to sleep.

"My grandfather believed in feeding everyone who came along and getting the latest news from them," says Dad. "He re-

> *"You can sleep in the barn, but remember, there's no smoking..."*

ally enjoyed talking with them."

At nightfall, he allowed them to sleep in the barn, Dad says, warning them, "Remember, no smoking."

Grandma Norah, however, had a different philosophy. "She believed everyone ought to be willing to work for a meal," Dad says. "She thought Grandpa was taken in by every freeloader that came along. But sometimes, work was well-received by the men and they'd stay several days to finish a job. In the summer, it might be helping hay or weeding potatoes. In the winter, strangers were usually set to cutting wood."

Dad says being a tramp was an accepted way of life for these men. "Their tales of riding railroad cars, being places we had never been and their carefree irresponsible life had considerable appeal for my little ears," he recalls. "But your Grandma managed to present a more realistic view of such a life when we were alone and squelched any ambition of the younger generation to follow suit."

Boarders, Too

With the Block School only 1 mile away, the Lessiters frequently boarded school teachers. In fact, Grandma Norah prided herself on playing matchmaker for school teachers on several occasions.

"Herb Lewis used to come out and go fishing a lot," recalls Dad. "Lulu Hammond from Clar-

---

The Steel-Wheeled Tricycle...

Dad used to keep a steel-wheeled tricycle which was handed down through several generations at his Grandma Wiser's place in Oxford so he could ride it on the village sidewalks.

"Sometimes, I would ride the tricycle to the store to get groceries for my Grandmother," he recalls. "Once I rode it to the store, forgot about it and walked home with the groceries. When I realized what I had done, I ran back to the store as fast as I could to see if it was still there. It was under the vegetable table where I had left it."

The trike survived one more generation, as I also learned to ride it in Grandma Norah's big house. I still remember those steel wheels making tracks on the rugs, but it was sure easier to ride there than outside in the dirt, mud and snow.

"I was really surprised your Grandmother let you ride the trike on her rugs," recalls Dad. "But then there really wasn't much she wouldn't let her only grandson do."

It was a sad day when I left the tricycle out by the milkhouse one afternoon where it couldn't easily be seen and Mom drove over it with the car. Since the steel wheels were too bent to straighten, my kids never had a chance to ride the old steel-wheeled trike.

kston was the teacher at the Block School and boarded at our house. Your Grandma matched them up and they got married.

"Lulu wasn't allowed to entertain boyfriends at home. But my Mother let her have boyfriends here anytime she wanted."

One woman was later hired to take care of Grandma Wiser in her later years, but they couldn't get along, so she stayed anyway and paid for her board. Other women worked for their board by cooking and cleaning.

Besides a married farmhand living in the tenant house, there were often one or two single hired farmhands who boarded at the big house during the 1920s and 1930s. In addition, there was usually a hired girl who sometimes lived in the neighborhood, but more often than not "lived in."

I am sure Grandma needed all the help she could get for washing, ironing and cooking. In those days of the wash board, carrying water, drying clothes outdoors, heating water and irons on the old cookstove, a woman's work was truly never done.

A Beautiful Home

There was some beautiful, hand-crafted furniture in that house. The dining room table could seat 20 and I remember a foot-pumped organ I never saw anyone play except visiting kids.

We eventually auctioned off everything, and if we owned that antique furniture today, it would be worth a fortune.

It was in that house where, as a kid, I raced Dad's steel-wheeled trike all over the rooms—you could really get it going fast. Later, my photojournalism career got its start in that house after I set up a darkroom in the big kitchen.

I also had a pool table in one of the vacant bedrooms, where my friends and I perfected our "three-corner jimmy" shots. I also tended the fire every day for my widowed grandmother.

All Gone Now

The house stood empty until it was sold. In late 1964, new owners spent thousands of dollars renovating the house. They never saw their work completed,

THE 501 HOUSE. Built in 1938, this was the new home for city girl Mom and Dad. It was located only 75 yards north of the barns.

"It was a very sad day when the big house and all its family history burned to the ground..."

however. Two young boys were playing with matches one day by the garage and the fire got away from them. The house burned to the ground.

Even though I had already left Lake Orion and the house had new owners when I heard about the fire, it was a sad day for me.

Deep down, I always hoped someday I could return with my wife and kids and live on Lohill Farm in that old house my Great Grandfather built.

New Home At Lohill

Being a city girl, Mom didn't really want to live out on the farm. Knowing she was making a sacrifice, Dad wanted to build something nice for his new bride. Grandpa Frank wasn't too happy about it—he thought the young newlyweds were going a little overboard.

They took out a $1,000 loan from Orion State Bank—where Grandpa Frank was on the board of directors.

"He begged us a number of times to pay off the loan," Dad says. "He was always embarrassed when they read the list of loans due at the board meetings and our names were called."

Dad says he kept asking them to pay the loan off. At one point, Grandpa Frank even offered to give my folks the money to pay it off. But they held off since the loan wasn't even yet due.

Another sore point with Grandpa was the colored bath-

room fixtures Mom wanted for the upstairs bathroom.

"Your Mother argued with me about the benefits of the colored bathroom fixtures," says Dad, "and your Grandfather would argue about what a waste of money they'd be. I was caught in the middle and didn't want to make either one of them mad.

"Finally, she said, 'Well, who is going to live in the house? Your Dad or me?' That's when I ordered the colored bathroom fixtures."

The fixtures weren't the only problem in building the house. After it was nearly completed, they began drilling a well. They drilled all around the house—even going 150-feet deep—and came up with nothing. Finally, after serious panicking, they located water at a shallow depth 50 yards away by the big barn.

Scary Housewarming

My parents moved into the new house on Halloween day, 1938—3 1/2 months after they were married. "We had our six-couple bridge club coming that night for dinner at 7 p.m.," Dad says. "We drove to Pontiac that afternoon and bought a big radio so we'd have some music while everyone played cards."

Just 90 minutes before their guests arrived, they didn't have any furniture. "We didn't know what to do, but the furniture finally arrived at 6 p.m. and we quickly set it up," he says.

Our house was nothing fancy, just a nice, comfortable house.

TIMES GONE BY. This old rocker on the front porch of the original farm house signifies quieter times long gone by. Instead of rocking on the front porch, folks today watch television and don't always get the stress out of their lives.

Painted white with green trim, it provided a serene foreground for our towering white barn.

As you walked in the back door to the house, you saw the kitchen and breakfast nook, which also had a door out to the garage. You wound around to the narrow dining room, which was separated from the parlor by the stairs to the bedrooms.

Once you reached the top, my bedroom was on the right and Janet's large bedroom was farther along on the right. The bathroom (with those colored fixtures), separated Janet's room from my parent's big bedroom, which equaled the whole width of the house on one side.

Downstairs was the green-carpeted parlor, where we spent most of our non-sleeping time. We played the piano, read books, listened to the radio and Mom tended to her cherished collection on the "what-not shelf" in that room.

Behind the fireplace in another small room was Dad's farm office, desk and typewriter.

One of the neat things about our house was the screened-in porch off the parlor that overlooked the barns. On summer nights, my parents sat out there drinking lemonade while the cars on Baldwin Road whizzed by. When it was really hot, Dad or I slept out there, searching for relief from the heat.

Different But Same?

We still get back to Lohill several times a year to visit Dad and my sister's family.

Every time we leave for Michigan, my wife chuckles when I say I'm going "home," pointing out I've lived over two-thirds of my life away from the farm.

She's right, but there are a lot of things about Lohill Farm I never want to forget. Things like my mom's chocolate chip cookies. Saddling up a horse and trotting back to the lake. Milking the cows at 5 a.m. with Dad. Playing hockey on the frozen swamp. Swinging on the tire swing at the oak tree.

A home isn't a building, a town, a quiet white farmhouse or 180 of the most beautiful acres you've ever seen. It's wherever you share good times with people you care about.

A lucky guy, I am. I have two.

Beef To Lamb To Milk To Eggs

IN THE early days at Lohill Farm, registered Shorthorn cattle and Shropshire sheep were seen grazing the fields.

By the late 1800s and early 1900s, the farm had developed a herd of about a dozen outstanding Shorthorn cows along with assorted bulls, yearlings and calves. The goal was to develop outstanding breeding lines so bulls could be marketed.

Success in breeding classes and capturing grand champion steer honors several times in the early 1900s definitely helped the sale of breeding stock.

"I remember one old Shorthorn cow that had horns that curved down and back to her eyes," says Dad. "We had to keep sawing them off to keep them from hitting her in the eye.

"She was a cow that wasn't much to look at from a conformation standpoint, but she always produced a good calf and was an important part of the herd for many years."

Meeting Milk Demand

But when things got tough in the beef business in the early

MORE COWS, LESS WORK. Shifting from a 20-stanchion barn to loose housing and a milking parlor let Dad double herd size without adding more labor.

1920s, the farm switched to a dairy operation. The trend at the time among farmers in southeast Michigan was to produce milk for the Detroit market.

The farm began selling milk to a Pontiac creamery in 1923. The mixed-breed-grade herd was housed in big wooden stalls in the big barn in those days.

Later on, family members tore out the wooden stalls and replaced them with used steel stanchions which came out of a dairy barn near Bloomfield Hills, Michigan. Dad remembers it was plenty of hard work installing the stanchions, pouring concrete, rebuilding the barn's support posts and putting in an overhead manure litter track.

A Super Option

The family started selling milk to Lakefield Farms in 1934, which was a good deal since there were no hauling costs for the milk. Each day, the Lakefield crew sent a truck to pick up the milk cans and paid the going market price for milk without discounts of any kind.

They processed the milk in their own dairy plant and hauled it to the J.L. Hudson department store and Harper Hospital in downtown Detroit. We continued to sell our milk this way until they dispersed the herd in 1956.

When we went back to selling milk to a plant in Pontiac, we realized how spoiled we had been for the previous 22 years since we were again hit with hauling costs, butterfat and quality discounts.

For years, the old milkhouse sat in the middle of the farmyard right next to the old, abandoned well. Cans of milk were kept cool in deep tanks filled with cold water. After an electric milk cooler that could hold four 10-gallon cans was purchased, cans were switched back and forth from the cooler to the deep tank to keep the milk cool in hot summer weather. Eventually, a modern milkhouse was built next to the big barn—which meant nobody had to lug those heavy cans filled with milk 150 feet uphill through winter ice and snow.

War Brought Changes

Dad was working on the farm and also doing plenty of electrical work wiring houses and local stores when World War II broke out. But because of the war-declared shortages, it soon became almost impossible to buy electrical equipment.

In 1943, Dad started a World War II defense job working on the Bofor gun at a Pontiac Motors factory. After just three weeks of factory work, he had to quit when Grandpa Frank developed heart trouble and the hired man left for another job. Soon after, Dad took over management of the farm.

Dad and Grandpa Frank gradually sold off all the grade dairy cows and started a purebred registered Holstein herd.

"We eventually had all registered Holsteins and were enrolled in the Dairy Herd Improvement Association milk testing program," recalls Dad. "We also enjoyed many sales of Holstein breeding stock over the years."

Since he needed a prefix to register the Holsteins, he came up with the Lohill Farm name.

"This was much to your mother's disgust as she didn't like the name," says Dad. "But the Holstein association accepted it and the farm name was accepted by the Michigan State Department of Agriculture.

"I still remember when you wondered why that name had been selected with all of the steep hills on the farm. I guess I selected it because it was different and a name that hadn't previously been used."

In 1945, the decision was

NOT SO PRETTY. Mom always prefered the cleanliness of the previous stanchion barn to loose housing where cows were free to come and go from feed, water and rest areas as they wished.

3,000 PLUS CHICKENS. Day-old chicks were brooded in the truck garage basement. Later, 2,500 hens would be housed in barns where manure dropped through slats suspended 18 inches off the floor.

made to speed up the milking procedures. Until that time, it took forever to milk the cows as they seemed to let down their milk at their own convenience.

So a three-minute egg timer became a critical part of the milking program. The milking machine was left on each cow for only three minutes. Even though the cow wasn't milked out completely, the machine was removed when all of the sand ran out on the egg timer.

Within a few days time, the cows learned to let down all of their milk in about three minutes. This change not only sped up milking, but actually improved milk production for many cows. And the few cows that wouldn't accept the new milking philosophy took a trip to the Detroit Stockyards.

No More Sheep!

Another change made in the mid-1940s was the decision to get out of the sheep business.

"We used to keep over 100 ewes, lambs and rams," says Dad. "But I was too tall and my back couldn't take bending over to get the lambs to nurse. We also had too many cattle to look after at that time."

Besides feeding lambs from the farm's flock, Dad and Grandpa Frank bought young lambs at the Detroit Stockyards in the fall months that were shipped in by rail from Western states. These lambs would be fed part of the farm's corn crop until they reached market weight and were sold at the Detroit yards.

"Having all of those ewes and lambs to look after was too much work and we were also beginning to have trouble with neighborhood dogs running in our fields and killing sheep," says Dad. "So we sold off the sheep to concentrate on dairying.

"We did okay with feeding the lambs, but I wasn't cut out for the sheep business."

New Dairy Facilities

Besides the new milkhouse attached to the big barn, a new bull and heifer barn was added in 1950. Dad and the hired man cut logs in the woods and hauled them to a nearby sawmill to be cut. The lumber was used to build the new barn.

Leon McGrath put in the foundation and laid the concrete

First Farm Visit

One bitter cold winter day, a bus full of foreign students from Detroit's Wayne State University came to tour the farm.

For many of these students, it was the first time they had ever seen a modern farm.

The highlight of the trip was not the barns, the machinery or the livestock. Instead, it was a tour of the farmhouse and their fascination with one of the kid's band hats which was sitting on a bedroom dresser.

One student had to go to the bathroom and took her shoes off outside—cultural shock to several Lessiter family members who had tracked plenty of mud across Mom's nice clean kitchen floor for many years.

Chickens, Chickens Everywhere....

....But Not a Single One to Eat

Chickens, Chickens Everywhere

Here's how a Detroit Free Press newspaper reporter wrote up the family egg operation in 1961.

JOHN LESSITER has 2,500 chickens clucking on his Oakland County farm. But when he wants one to eat, he finds it cheaper to go to the store and buy one from Georgia.

Lessiter's chickens—egg layers—cost 45 cents each when they're only 1 day old and weigh only an ounce. But in Detroit-area chain stores recently, fully dressed broilers have sold for as low as 19 cents a pound.

As old fashioned as a buggy whip is the combined operation of a few hundred birds with eggs yielding the farmer a small but steady profit and the meat netting him 35 to 50 cents per bird.

High Chick Costs

Lessiter's chick costs are high because of the millions of dollars poured into research to develop hens that will average 93-percent production (a little over 27 eggs in a 30-day month). But the slow-growing strain is so worthless for meat purposes that the hatchery kills all the males as soon as they're hatched.

Lessiter, whose Oakland County operation is a model of mechanization, increases his net profit

> *"The farm increases net profit margin by packing and marketing high quality eggs..."*

margin because he packages and markets his own product.

"In the old days," says Lessiter, "egg production amounting to only one-quarter of today's was not uncommon. People used to have straw on the floor and let chickens scratch in it to find whole grain and oats to eat."

The commercial feed for his white Leghorns consists of a mixture of corn, oats, soybean meal and antibiotics. It is carried on a flat chain inside a three-inch-wide trough that runs back and forth in a serpentine pattern through the whole area of the chicken houses.

The machine is timed so the chain runs every 120 minutes during the 14-hour day, then shuts off for the night.

Also provided are water troughs and containers of ground oyster shells to reinforce the diet with calcium and other minerals.

Lessiter's 2,500 chickens eat continually during a standard 14-hour day—maintained artificially with electric lights in the winter.

Lessiter keeps his 2,500 egg layers high and dry—confined in houses with slatted floors raised a few feet from the ground.

The laying nests, with beds of ground corn cobs to keep the eggs from cracking, line the walls.

A Hot Start

Lessiter must keep baby chicks in a brooder house heated at 90 degrees F for the first week. Then he can gradually reduce the heat over eight to 10 weeks. The chickens are four and a half months old before they start to lay eggs.

Plenty Of Eggs

The farm produces 1,000 dozen eggs a week and Lessiter has calculated that each dozen eggs takes 4 pounds of feed at 3.75 cents a pound.

Eggs are picked up from the nests every hour in baskets and are sorted in a mechanical egg grader that weighs each egg and rolls it into compartments for packing in extra large, large, medium and small size egg cartons.

Produce—Or Leave!

To maintain constant production, Lessiter must keep chickens of varying ages in different houses.

They continue to lay at a profitable rate until they are 20 to 24 months old. But as soon as they fall to 50 percent production—one egg every other day—Lessiter places an advertisement in the local newspaper and sells them live as stewing chickens at 50 cents apiece.

—*Harry Golden Jr., in the July 2, 1961, Detroit Free Press*

block for the walls.

Laminated rafters for the upstairs hay storage area were shipped via railroad boxcar from the manufacturer to the siding at the Lake Orion Lumber Yard. These rafters were hauled home on a trailer behind the pickup truck and pieces were assembled on the second floor of the new barn and winched into place.

Still More Cows

In 1955, we made a switch from a stanchion barn to a loose housing facility with a three-stall milking parlor. The family started milking more cows and using less labor, but it was touch-and-go in the dairy business back in the mid-1950s.

Mom always believed the loose housing setup was dirty with manure splattered all over the place. She remembered how clean the 20-stanchion barn had been with white lime spread evenly on the alley floors.

She was right. But the stanchion setup was a great deal more work and we were limited to milking 20 cows.

I remember the first night we milked in the parlor. Mom maintained the cows would never walk up those steps and through the door to be milked in the parlor, but most did so without too much pushing and shoving.

They were skittish and didn't give too much milk that night. While a few cows had to be led into the parlor to be milked on opening night, all the cows soon got used to the milking parlor and production went up sharply.

By feeding grain only in the milking parlor, there was a real incentive for the cows to line up eagerly to be milked.

One time, a cow trapped Dad against the concrete block wall in the rear of the parlor. Somehow the cow had gotten turned around in one of the milking stalls and Dad was trying to free the panicked cow. All of a sudden, she lunged forward and Dad was trapped between the cow and the wall.

For what seemed like a long time, neither Dad nor the cow could move. He could hardly breathe. Somehow Dad got free, but he had sore ribs for more than a week.

Because of poor milk prices, we decided in 1958 to sell the herd. A northern Michigan farmer trucked them to their new home.

From Cans To Cartons

That was the year the farm changed over to egg production. We sold eggs to schools, grocery stores, camps and about 20 Lake Orion families who had been on our two weekly egg routes for many years. Dad found he could handle the egg work by himself without having to hire extra labor—or so he thought.

"The bottom had fallen out of the dairy business and we had to do something different" says Dad. "We found producing eggs was much more profitable than the cow business had been.

"We had good egg production, good sales and a good market where we got retail prices for almost all of the eggs. But we worked hard to develop that market. We had a good quality product and good packaging which helped."

At one time, the farm had 2,500 hens housed on 18-inch-high slatted floors in the big barn and in the former heifer and bull barn. Dad and Mom started out washing and candling eggs in the basement of the house, but moved this operation to the milkhouse after the dairy herd was sold.

A refrigerated room was built in the milkhouse. Once they were gathered, the eggs were washed and then cooled before grading. This procedure definitely helped keep the eggs fresh and most eggs were sold the day after they were laid.

But if Dad thought dairying took plenty of labor, he soon found that egg production required even more time. While the work wasn't as taxing as feeding, milking and cleaning up after the cows, someone had to be there to gather eggs at least a half-dozen times daily seven days a week.

After Dad became township supervisor in 1961, he cut back on farming and eventually leased out the egg business. But that didn't work out very well and the family finally sold the egg business a few years later.

BASEMENT EGG ROOM. When the farm moved into egg production in a big way, Mom and Dad graded and packaged eggs in the basement of the house. Once the cows were sold, the egg processing moved to the old milkhouse which had a refrigerated cooling area.

Livestock Tales Of The Past

FOR MORE THAN 100 years, livestock played a key role in our family's farming operations. Here are a few stories which family members recall from the past.

Horse Tales

Dad remembers a pretty team of light Mustang horses. Tough to harness, they always wanted to run once you got them hitched to a wagon.

"They ran away a number of times," he says. "They were dapple gray in color, so they must have had a little Percheron blood in them. These lightweight horses had been raised out west on the open range."

There was also the team of

OLD FENCE POSTS. These worn-out fences and posts tell the story of earlier days when beef cattle, dairy cattle, hogs, sheep and horses roamed the many fields at Lohill Farm.

123

HOGS WERE DIVERSIFICATION. While the farm never had large numbers of hogs, a number of sows were farrowed annually in A-frames for a number of years.

Belgian horses which were plenty creative when it came to figuring out ways to escape the barnyard.

"In the middle of the night, we'd be asleep and hear horses running through the front yard or out on the road," says Dad. "It would always be this darned team of Belgians.

"It didn't matter what we did with the gate—somehow they would always open it. Then we'd have to get up and chase the team down the road in the middle of the night. And the cows would get out too."

There was also the horse which the veterinarian decided was actually allergic to hay. He and another Belgian were used on the farm to rake hay and he wheezed all the time when he was working in the hay field.

Finally, it got so bad he had to stop and rest every 5 minutes, so we had to sell him. And that was the day that brought an end to the era of using real horses on the farm back in the early 1950s.

Cow Tales

Dad recalls the pitch-dark nights on the farm when it seemed like you couldn't see anything in front of your face.

"We had a bunch of black cows that would get loose and

"Chasing black cows on the blackiest of nights could be very scary..."

we would have to chase them," he recalls. "We'd find ourselves right on top of them and we still couldn't see them because of their black color and the moonless night. This scared the daylights out of me!"

When I was growing up on the farm, pasture ground was always in short supply due to our limited farm acreage. As a result, we pastured cattle in several other places.

One place is now the Clarkston Golf Course where we started pasturing yearling Holstein heifers in 1946. This pasture had been an old golf course which had gone broke during the Depression and the cattle grazed the tees, fairways and greens.

A contractor had purchased the course during World War II, dug up all of the old water pipes and sold them for scrap.

He later turned this land back into a golf course in the mid-1950s, so we had to find other land to pasture heifers during the summer.

We used to truck the yearling heifers over to the old golf course in early May and bring them home in September or October after the grazing season was over.

We usually rounded up the

THE OLD TREE. Located halfway down the farm lane, this old tree served as a handy backrub for thousands of cattle over the years.

A CHANGE IN BREEDS. While Shorthorns ruled the farm in earlier days, the 1970s brought Hereford, Angus, Charolais, Simmental, Maine Anjou and other exotic breed crosses to the farm as Janet and Neil Roberts put together a small herd of excellent beef cattle.

hogs and sheep up and down Baldwin Road since traffic was rarely a problem in the old days.

Rented State Ground

When I was a kid, we also rented pasture ground for a few years from the state of Michigan at a location a few miles southeast of Lake Orion. This ground later became a state park and was plenty rugged and hilly.

In fact, the brush was so dense in some areas that we could hear the heifers moving a few yards in front of us, even though we couldn't always see them.

One time when I was a kid and helping put out salt and check the heifers, Dad took time to show me an old cabin on the state land.

Years later, this old cabin made the national news as the place where suicide doctor Dr. Jack Kevorkian helped one of his patients end her life.

heifers and got them into the truck for the short ride home. But there was one time when there was one heifer we just couldn't seem to catch.

After she had been running loose on some of the hilly Lakefield Farms pasture ground behind the lake, it looked like we had her cornered. But somehow she managed to break free and swam the nearly half-mile distance across the lake.

Something must have really frightened her because she started to act crazy. Later, her head started to swell up and she had some health problems. We soon had to sell her for beef as nobody could control her.

Dad says the family also used to rent a few nearby fields and muck ground for needed pasture. It wasn't unusual to drive cattle,

WHERE THERE'S LIVESTOCK. The old horse-drawn spreader worked as hard as any implement on the farm. Pulled by Silver and Scout, the veteran Belgians that made their home on the farm, manure was spread on almost a daily basis.

There was also the time when I was getting a bred yearling Holstein heifer ready to show and Dad didn't want to pull her down with the rope sling to trim her feet. Instead, I walked her a mile over to the main Lakefield Farms dairy barns so Kent Mattson could trim her in the stocks.

By the time I'd walked her home, her hooves were in perfect shape.

Frank's Tale

Over the years, I've heard Dad tell our four children many times how tough it would often be to get me up on a cold winter morning to help with the milking and chores before I headed off to high school each day.

When Dad got up at 5:45 each morning, he'd stop in my bedroom and wake me just before he went down to the basement to put on his barn clothes.

If I still wasn't up a few minutes later when he was going out the kitchen door to the barn, he'd again yell for me to get up. If I still hadn't gotten up 15 minutes later and he was in the middle of milking, he'd run to the house and really yell at me.

That didn't happen very often. But when it did, I knew it was definitely time to get moving!

On bitter cold winter days, I'd sometimes make it down to the living room where I would curl up in a ball in front of the hot air register to try to warm up for a few minutes before getting dressed and heading out to the barn.

Sleeping In The Silage

Still another trick I used to warm myself on a below zero day was to climb the silo chute which I had to do to toss down silage for the cows. I'd quickly fork down enough silage to dig a 2-foot deep hole. Then I would crawl into this hole and and let the heat escaping from the newly dug silage hole warm me up.

Occasionally, I'd fall asleep for a few minutes—nice and warm and contented.

But I'd come awake in a hurry when Dad realized I wasn't working and yelled up the silo chute where his voice echoed all over the place.

THREE YEARS OF WINS. The Oakland County Livestock Judging Team took top honors in 1936, 1937 and 1938 during the annual mid-winter Farmer's Week contest at Michigan State College. Making up the 1938 team were Dad (seated at far left), Graham Bodwell, K.D. Bailey and Albert Kessler.

The 3-year reign led to permanent possession of the trophy and these consecutive wins were especially noteworthy since the rules required a different team to compete each year.

Chicken Tales From Lohill Farm

A FLOCK OF laying hens always played an important role in our family farming operation from the time I was a youngster until I headed off to college.

We had a small flock of hens which supplied the three farm families with eggs. Dad also had a small Friday night and Sunday morning egg delivery route in Lake Orion.

When I was a kid, Dad would order two boxes of day-old female chicks each January. An old quonset-roofed-style brooder house on skids was pulled up close to the house so we could keep a close eye on these 100 chicks during the extremely cold January and February weather. After the chicks got a little bigger, it would be pulled out into a pasture field and the chickens would be allowed to run outside during the day.

After School Job

While I fed and watered these young chicks, my job after a day of elementary school was to feed, water and gather eggs from the 50 or so laying hens housed in the old chicken coop.

In later years, the laying flock was expanded to as many as 3,000 birds. Even with automatic feeders and waterers, someone had to be on hand all day to collect, candle, wash and pack eggs. Production would be around 200 dozen eggs a day.

With this much bigger flock, the potato cellar was remodeled for starting out day-old chicks. This meant lugging pails of fresh water from the old milk house, carrying them 75 yards to the former truck garage and walking down a steep set of stairs to where the young chicks were housed. It also meant carrying

EGGS AND MORE EGGS. With as many as 3,000 laying hens on the farm, there was plenty of work grading and packing eggs in cases for sale to outlets such as the local school cafeterias.

50-pound sacks of chick starter down those steep stairs.

Dark, But Light?

As the birds became a few weeks older, we would replace the light bulbs with dark blue bulbs so we could easily catch the chicks for debeaking. This prevented them from wasting feed and sharply reduced the injuries from the birds pecking one another. This was tough, dusty work with 1,000 chickens to do at a time and we usually did it twice while they were pullets.

The dark bulbs let the farm crew see what it was doing, but the colorblind chicks remained calm, thinking it was nighttime.

As the chickens grew older and were ready to start laying eggs, they were transferred to one of two 1,500 bird slatted-floor units in the old dairy barn.

With these slatted floors, manure dropped into piles under the floor and we only pulled out the slatted floor sections and hauled manure every two years.

Two stories come to mind when I recall these large flocks of laying hens.

The first one I remember is when Dad had gathered a full basket of eggs and fell through the slatted floor into the dry manure pile located 18 inches underneath.

While falling through the floor was certainly a big shock to him, Dad didn't get hurt. But most of the wire basket full of eggs were broken. We didn't fill egg baskets that full after this happened.

The other story deals with the bird that became known as the farm's "Super Rooster."

One year when we ordered 1,000 sexed day-old female birds, we got only one male in the boxes. Usually, we would get eight or ten males since it's not easy for even a person with a highly trained eye to sex these chicks as they are born. As a result, an occasional mistake was made.

Since there was only one male bird, Dad allowed him to run with the 999 females. And he was still a member of the flock when the birds were transferred to the laying house.

Actually, it became a continuous joke as the farm crew imagined how this sly old rooster was getting along with all those

"We called him the Super Rooster as he ran with 999 hens..."

females. Everyone figured he was really enjoying life.

But living with those 999 females soon brought a big change in his life. It even got to the point where "Super Rooster" could be found sitting on eggs in the nest and acting like a laying hen himself. What a bird!

Rotten Eggs!

As a teenager, it was fun to find a few eggs which had fallen through the slats and were sitting unbroken on the manure pile.

More than once, my sister and I had a few good egg fights—and you can't believe the horrible sulfur smell a six-month-old egg gives off when it splatters off your shirt.

One time, Neil and a friend filled his bright yellow pickup with manure from the hen house. After the friend unloaded the manure, he discovered it had eaten the paint off and he had to have the truck repainted.

City-Reared Kids

There's one other chicken tale I want to tell you, but this one didn't happen until our kids were growing up. The family was back at the farm for one of our numerous visits with the grandparents and my sister's family.

I'd grown up often enjoying a Sunday afternoon dinner of fresh baked chicken which had been a live bird until late Saturday afternoon. But our kids had never imagined what it took to get a chicken from the henhouse to the dinner table.

So you can probably imagine the flack I received from my city-reared wife when the kids ran in and told Mom how they'd seen chickens dancing around after their heads had been cut off.

To a country boy who had seen this for years, it was no big deal. But to a city-reared kid, it was a traumatic experience that they'd never forget.

I just wish they'd never told their Mom!

WIRE BASKETS. For years, members of the Lessiter family collected eggs each day in these old wire egg baskets.

Memories Of Real Horsepower

IT'S BEEN MORE than 45 years, but I still remember when we farmed part of the home farm with Belgian horses.

While we were pretty well mechanized with tractors by the time I came along, we still used one team of Belgians to do some of the farm work.

The most horses ever used on our family farming operation that dates back to 1853 were three teams of horses and a driving horse that had been a wedding present for my grandparents. Anyway, by my time on the farm, we only had seven horse stalls in our big dairy barn and some were used to house dairy cattle.

When I was a kid, we had a team of big Belgians named Silver and Scout. I guess it was okay to borrow the names of the horses which the Lone Ranger and Tonto rode to radio and television fame since the originator of the Lone Ranger radio series, Brace Beemer, lived only 3 miles away from our family farm.

Known For Slow Pace

The team was the domain of our hired man at the time, Harry Robertson. I'd swear those two horses often took on the same personality and movements of

LONG AGO TIMES. Memories of the old Belgians still remain with some of the neck yokes and other items left from the days when horses ruled the farm.

PLOW POWER. Grandpa Lessiter explains to grandson, Mike, how the Belgian horses pulled this moldboard plow in the fields.

old Harry. In fact, they moved the same way he did—at a fairly slow, steady pace.

Besides that, this team was constantly opening the barnyard gate at the worst possible times. Those two horses could nuzzle that gate open with ease—practically anytime they felt like doing so.

I can vividly remember how my dad hated to hear the clop-

DISKING WAS SLOW. With this small disk pulled by the team of Belgian horses, it took a long time to work fields. The steel top frame held heavy rock when the ground was hard and more weight was needed on the disk for essential penetration.

CLOCKWISE FROM TOP LEFT: An old singletree used with the farm's team of Belgians, trace hook on a singletree, clevis-connected singletrees, singletree hitching ring, bridle and a lead line used over the years at Lohill farm.

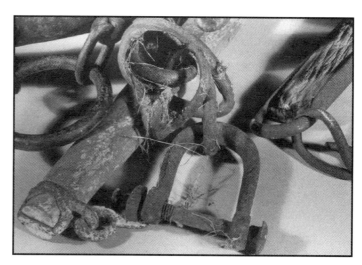

YEARS OF MOWING HAY. A mower like this was used for years to mow hay at Lohill Farm. Then came a day when we cut the tongue, retired the horses and started using the tractor. A youngster rode the old steel mower seat to lift the knives at the end of the field. Dump rakes, such as the one shown at right, were fashionable for years before being replaced by much more efficient side delivery rakes.

clop of their hoofs out on the blacktop road in the middle of the night. He especially disliked getting up on cold winter nights to chase them down, to return them to the barn and to close the gate.

LOOSE HAY HANDLING. Before balers came to the farm, hay was cut, raked and elevated directly from the windrow with this hay loader to a wagon drawn by horses. Two men with forks would spread the hay on the wagon before heading to the barn with another full load.

pace with the horses pulling the hay wagon back to the field after we unloaded since it gave us all a much-needed chance to rest for a few minutes.

He was willing to grant that lull in haying if the field was fairly close to the barn. If we were working in one of the back

---

"I remember the day when the horses were retired from cutting hay..."

---

fields or if it looked like the hay would soon be getting wet, Dad would grab the lines and Silver and Scout would be off at a fast trot.

We used horses in those days for mowing hay, raking hay, hauling hay, hauling manure, pulling the grain binder, cultivating and even for pulling the tractors and truck out of the mud.

Time To Change

For years, we mowed and raked all our hay with horses. But eventually there came a time when Dad realized this was slowing down the haying season too much.

I still remember when he took a saw to the tongue of the hay mower and cut it in half. After that, we mowed with the tractor. As a young kid, I spent many

But I'll give Silver and Scout credit for one thing. They were smart. On those occasions when my dad picked up the lines, they immediately perked up their ears and were ready to move. They always knew when he grabbed the lines...and they quickly realized they had better get moving. If they didn't, they knew the end of the lines would soon be applied to their rumps.

Dad was normally willing to let Harry plod along at a slow

summer days riding the mower seat behind the tractor. My job was to raise the sickle bar at the end of the hay field or when it became plugged with hay.

Silver and Scout were used mainly to rake and haul hay. That was a nice slow job suited to both the horses.

Allergic To Hay

But then one day Scout got sick while raking hay. Raking wasn't such a tedious job...and should have been something any horse could handle.

After a close inspection, our veterinarian said the horse was allergic to hay. Now that's really something—an animal who eats and works mainly with hay actually being allergic to it.

Since this happened during

> *"The vet said he thought old Scout was actually allergic to hay..."*

the peak of the haying season one summer, the easy solution was to get out the saw and shorten the wooden tongue on the rake. After all, just one horse on the rake wasn't very efficient.

And that's just what we did. So hay raking became another "tractor power" job on our farm.

And it wasn't long after this when we said goodbye to the team as they were sold to another farmer.

Rather than being allergic to hay, I still believe old Scout was mainly allergic to hard work.

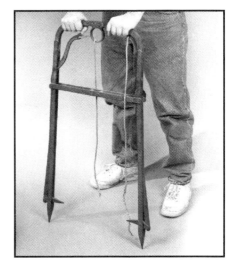

ROPES, PULLEYS, HAY KNIVES AND HORSES. These pulleys, long ropes and hay forks were used to effectively move loose hay from the farm wagons to any one of four mows in the big barn. First, the horses would pull the wagon load of hay up the hill into the big barn. They would then be unhitched from the wagon, be given a drink of water and hooked to a set of whipple trees on a long rope. After the hay forks, right, were set into the hay on the wagon, the team of horses would walk 30 yards down the hill which would in turn pull the hay via ropes and pulleys to the top of the barn where it could be pulled with another rope by the farm crew to one of the back mows and be dumped. Each wagon load of hay usually took three trips to unload.

The Old Farm Lane

THE HOME FARM was laid out in a fairly narrow L-shape that stretched east from Baldwin Road and then wrapped itself around three sides of Dark Lake.

From the 1930s on, 14 fields made up the 175-acre farm.

A fenced-in permanent farm lane split 10 fields which were located east of the barns. Wrapping around the lake, the other four fields each had a semipermanent, usually unplowed wheel track path running through the fields which served as an unofficial farm lane without any fences.

A number of trees grew along the fenced-in farm lane. Several trees replaced posts as a place to drive staples to hold the woven wire and barbed wire fence secure. All of the fields were fenced with gates to allow livestock to be pastured.

A choke cherry tree had a low limb where a child could sit until the cows went by.

During the summer months, the Holstein cows would usually spend the day in one of these pastures to graze and cool off in the shade. Then they would be driven back in late afternoon to the barn for milking.

After being fed and watered, they would be pastured overnight in a field located closer to the barns or left in the barnyard so it didn't take but a few minutes to get them into the barn for the morning milking.

Although there were a couple of low spots in the farm lane in

FEW LANE CHANGES. Nobody knows how many years ago this photo was taken, but it certainly shows the old Lessiter family farm lane hasn't changed much in many decades. For many years, it split the farm down the middle and made it easy to move horses and implements to fields.

which someone would occasionally get stuck with a tractor and manure spreader, baler or a wagon load of hay, it was pretty much an all-weather surface road that could be used most of the year.

Haying Wasn't Always What It Seemed To Be

ON OUR DAIRY FARM, haying season always seemed to run almost non-stop from early June to early September.

Of course, the actual length of the haying season usually depended on the weather and whether there was sufficient moisture for a late summer third cutting.

I won't bore you with all of the details of putting away a big crop of hay to feed our dairy cows through the winter. But I want to relate a few interesting tales which made haying exciting for me over the years at Lohill Farm

The Big Motor

In the days when we were still putting up loose hay, we installed a slotted-floor hay drying system in the big barn.

The fan was powered by a 7 1/2-horsepower electric motor. That was really a lot of electricity in those days—definitely the

YEARS OF HARD WEAR. Hinged to keep cool drafts from blowing down on the cows, this hay mow door was used for many decades to drop hay down to the dairy cattle stanchioned below.

THE BACK 40. Some of the back fields were very rough and hilly, but the normal alfalfa and corn rotation was still followed. It could get pretty scary harvesting a hay crop off these steep hills.

biggest electric motor we used on the farm. But it allowed us to cut hay at higher moisture, dry it in the mows and produce higher quality forage.

For 10 months of the year, we'd use this motor to grind feed

"In the summer, the electric bill was always higher at the barn than the house..."

almost every day for the cows. During June and July when we were using the motor to dry hay and the feedmill was without electrical power, we'd head to the elevator in nearby Oxford to have corn, oats and wheat ground for the milking herd, heifers and calves.

During haying season, the electric bill at the barn was al-

HAY TREAT. When the grandkids came to visit from Wisconsin in mid-summer, it was always fun to help Grandpa Lessiter haul hay. The big barn's banked hill led up to the second floor so a wagon filled with bales could be driven right into the hay mow area for easy unloading.

ways much higher than the bill for the house. The hay drying fans ran 24 hours a day until the hay was dry.

One time some of the officials from the Detroit Edison Co. even

"The quality of the old hay was terrible, but it sold for a premium price..."

came out to see why the utility bill was running so high.

3 Weeks Of Non-Stop Rain

Years later, the last field of first cutting hay was in a back-of-the-farm, 30-acre field. Every time it was dry enough to bale, it would rain again and then we'd have to wait for the crop to dry one more time. This musty smelling hay was in pretty sad shape after 3 weeks of lying in a wet, damp windrow.

By the time the first cutting was finally dry enough, it wasn't worth much since the quality had long since disappeared. Out of desperation, we stored those bales in the barn in hopes the second cutting would come on and we wouldn't lose too much of the alfalfa stand.

The quality was so poor that the hay sat unfed in a corner of

HAND GRASS SEEDER. Strapped over the shoulders, this grass seeder was used by several Lessiter generations to seed hay and pasture fields to orchardgrass, alfalfa, brome grass, rye and other forages. The key was to walk at a stready pace and know how fast to turn the crank to get an even seeding rate. Even in later years when these crops were drilled, this hand seeder was used to seed wet spots where a tractor might tend to get stuck.

the barn for 5 years. Dad knew there weren't many pounds of milk in those bales.

Then one winter, the two Thoroughbred racetracks in the Detroit area were desperately looking for anything to use as bedding as wood shavings were in very short supply.

Wouldn't you know it? Dad sold the old, low-quality hay for racetrack bedding and got more for it than the crop would ever have been worth as livestock feed.

Hay Bales And Dinner

One of only a few things that ever annoyed Mom was when Dad would jump up from the dinner table and head out to the barn to sell one more bale of hay. Folks knew they could stop at any time and likely find Dad there.

It always made Mom mad to see him jump up from the table, leaving a nice warm, delicious meal cooling on the table.

She felt these folks needed to learn not to show up during the dinner hour. And if just once Dad didn't rush out there or told them to come back at another time, then dinner wouldn't get cold, she maintained.

Hay Bales, Corn Flakes

Speaking of the dinner table, our four children have heard me tell many stories over the years about all the tough work there was to do on the family's dairy farm.

Sitting at our dinner table in the Milwaukee suburb of Brookfield, I'm sure they tired of me

"Being a long-time successful dairyman and knowing what forage and bedding was really worth, Dad could never bear to charge what people would pay for hay and straw..."

telling them how hard I'd worked as a kid while all they had to do was take the garbage out once a week.

One of the stories I told many times was how we handled 70-pound bales and how tough it would be by the end of the day to toss them up on the wagon or truck.

One time when our family was back at the farm, our son Mike, who was 11 years old at the time, was going to help haul hay. Remembering my story of the 70-pound bales, he reached over to pick up a bale out in the field and it came up so fast that it smacked him in the chin and he went sprawling on the ground.

While the rest of us had a good laugh, Mike didn't think it was funny. He suggested I'd been fibbing all these years about the heavy hay bales.

While I hadn't been fibbing about the 70-pound bales, I had forgotten to explain to him how his grandfather had been baling lighter weight bales every year.

That's because Dad always had a problem when it came to pricing hay for the suburban market. Being a dairy farmer, he had a limit as to how much he thought he could charge for a bale of hay—basing his pricing on the nutritional value a bale of

"What my Dad had always claimed were really heavy bales now only weighed 35-pounds..."

hay had for a high-producing dairy cow.

When I'd go back home, I'd see what he was charging for hay and would tell him that wasn't enough. I tried to reason with him that hay was scarce in this area, that the folks buying a bale

PLENTY OF PROSPECTS. With lots of cars rushing by on busy Baldwin Road, there was plenty of hay and straw business at the Lessiter Centennial Farm. Note how bales of straw often sell for more than hay, something that was always difficult for a dairyman who had to buy hay for years to fathom.

143

A PROBLEM FOR MOM. When a truck or car horn blew at the barn, Dad would jump up from the dinner table and rush out to sell someone a bale of hay. It would always irritate Mom who believed people should not show up at the farm to buy hay or straw at odd times, especially at supper time. It was always a constant battle at the supper table which Mom never seemed to win. Dad believed the customer always came first.

or two at a time had the money and weren't really worried about the value of a ton of hay. But he never listened to me.

I guess he finally tired of listening to me preach about his low hay prices. He soon came up with the idea that if he didn't want to charge more per bale for the hay, then he'd bale lighter weight bales.

So the weight of a bale of hay gradually dropped to 60, then 50, then 40 and finally to 35-pounds. That's what had happened by the day Mike thought he was pick-

"His marketing of hay was very much like today's packaging of corn flakes..."

ing up a 70-pond bale and instead found a 35-pound one smacking him on the chin.

Dad's marketing strategy on turning out lighter hay bales is the same pricing concept breakfast food manufacturers follow when they package corn flakes.

You never find a full box of corn flakes when you open up a new package at the breakfast table. It was just Dad's way of keeping the price down!

Those Really Hot Days Of Summer

A FEW WEEKS BACK, our home's central air conditioner simply gave up. As a result, we suffered through a terrible week of hot, muggy weather until a replacement unit could be installed.

While trying to sleep in this 95-degree-plus weather, I got to remembering growing up at Lohill farm in the days when it always seemed like the summers were hot. Like most farm families, we didn't have an air-conditioning unit to get us through those exceptionally hot summer days and nights.

The temperatures often climbed close to 100 degrees and the nights didn't cool off. It often seemed like it was 110 degrees while trying to sleep in my upstairs bedroom. Even though the two windows were open, the heat made you toss and turn throughout the hot, hot night.

Unbearable Nights

When it got unbearable, Dad used to go downstairs and sleep on the squeaky glider in the screened-in porch. There were often cooling breezes which allowed Dad to sleep much better.

Yet the few times when I tried sleeping on the porch, I usually couldn't get to sleep or would

HOT SUNNY DAYS. When the days turned the hottest was when it always seemed like the hardest work had to be done on the farm. Such was the case at Lohill Farm.

continually wake up. The noises from the passing cars speeding by on the blacktop highway would keep me from sleeping. Or the glider which squeaked with almost any movement of my body kept me from dozing off.

Looking back, I don't remember why we never slept in the basement on those hot nights. That would have been the coolest place in the house to drift off to sleep on a steamy night.

Long, Hot Days

What would really get to you was when you got up at 6 a.m. to head to the barn for chores and the air was still, the sun was bright yellow and the thermometer was already approaching the 90-degree mark. With these early morning conditions, you always knew it was going to be another tough, miserable day.

While no farm work was ever fun on these sticky summer days, some tasks were worse than others. Whether it was handling loose hay or baled hay, it was a hard, dirty, dusty job with plenty of itchy hay leaves floating down on your sweaty skin.

Putting the hay away in the hot, stuffy mow was always worse than loading bales in the field. Yet the best part was when you had finished putting away another load of hay and climbed down the ladder to the barn floor.

Then the crew would cool off with the water hose and enjoy a nice, cold drink of water or a Pepsi. And you'd really enjoy the breeze as we headed back down the lane to the field to get another load of hay.

Shocking Work!

While haying was hard, sweaty work, shocking wheat, oats and barley could be even tougher. The grain was never ripe until the very hottest summer days of late July and it meant making round after round in the field to shock the grain so it could dry and be protected against the weather.

Unlike haying, you didn't get many breaks on those "shocking" days when grain was cut with the binder. We'd carry a jug of ice-cold water and leave it in the shade of a grain shock located at the start of each round.

Where Is It?

But sometimes you couldn't remember the particular grain shock where it was when you worked your way back around the field. You would be dying of thirst and it would often take a while to find the water jug in one of the shocks of grain.

I always maintained Dad used that cold-water jug as an incentive to keep the farm crew working hard when shocking grain.

Once the crew had worked halfway around the field, they started thinking about that ice-cold water and then worked all the harder to get back to it as quickly as they could.

CENTENNIAL FARM HONORS...

When the Lessiter family farm reached its 100th anniversary in 1953, there were only 393 officially registered Centennial farms in the state of Michigan.

The first Centennial Farm in Michigan was started around 1815. There are about a dozen Michigan farms which were started prior to 1831.

As of the mid-1990s, records show about 5,000 Michigan farms qualified for Centennial Farm status.

HAY TO BE HARVESTED. This early summer aerial shot of Lohill Farm shows plenty of hay bales to be hauled to the barn from this field next to the farm house and barns.

Threshers For Dinner— In 20 Minutes!

FIRST OF ALL, let me start out by telling you my mother was a city slicker. When she married Dad and moved to the farm, she knew little about agriculture or any of the novel country ways of farm folks.

But having made summer visits to her aunts' and uncles' farms in Canada during her childhood, Mom did know the many wonders of overflowing tables of food often prepared for mighty hungry threshing crews.

A Social Occasion

She remembered how neighbors for miles around always came to help out at these mid-summer noon-time gatherings of hungry, sweaty, hard-working men.

While threshing was always hard work, it also served as a social affair in many communities such as ours for farm wives during the busy summer months.

WHEAT WAS A BIG CROP. Because of higher yields, wheat was always the favorite small grain crop. But we also grew oats, rye and barley on occasion.

BUSY HANDS. Holding the reins for the team of horses in your hands, properly manipulating the controls on the grain binder and keeping a close eye on the cutting knives, the dropping grain bundles and much more always took plenty of keen concentration on the part of the farmer sitting on the old steel grain binder seat.

With the last bundle of wheat dropped on the ground, the binder was ready to move to the next field. This meant unhitching the team of horses, unbolting the tongue and swinging it 90 degrees to the position shown here and rebolting it. Next, you added a pair of easily removable steel wheels to pull the unit through the farm gates.

HOT SUMMER WORK. The neighborhood's teenagers always seemed to get the opportunity to shock grain so the bundles could dry out before threshing. It meant standing the bundles on end so any rain would fall off while hoping the whole grain shock wouldn't eventually topple over. Besides, it was hot, dirty work.

When you worked your way back around the field and desperately needed a cool drink of water, it seemed like you could never remember which shock of wheat you put the water jug in to cool.

Just a few weeks later, every one of these grain bundles had to be pitched onto a wagon to take to the barn for further drying or directly to the threshing machine.

DRY MOW STORAGE. To protect grain from the rain before the threshing crew arrived, we stored grain bundles in one of the dry mow floors of the big barn. Rope grain slings were laid out on wagons before loading in the field and bundles were later moved onto this pile stacked as high as the beams.

The threshing machine would be pulled into the next mow and the grain bundles would be fed directly into the machine. After all of this grain was threshed, several wagons overfilled with bundles of grain would be pulled into this area and fed into the roaring machine.

Soon after getting married and for several years afterward, Mom had helped Dad's mother feed the threshing crew several times.

Then...Her Turn Came!

With Grandma Lessiter getting up in years and operation of the farm shifting to the younger generation, there came a year when Mom and Dad decided it should really be her responsibility to feed the threshing crew. And for many years afterward, she vividly remembered her very first "in charge" threshing din-

> *"She always remembered her very first completely-in-charge threshing dinner..."*

ner—with plenty of good reason!

Yet it is to her credit that the men did not go away hungry. Nor was it the food she served

THRESHING TIME. A big old coal-fired, water guzzling steam engine used to pull the threshing machine to the farm in the early days. In later years, we relied on the threshing machine from the farm next door. We helped man the threshing crew at their place and their men later came down to our farm to get the small grain threshed and into the bins.

which gave her a serious case of nervous indigestion.

As the story goes, the men were threshing at the Frank and Neal Dowling farm located a half mile down the road from our place. When Dad returned home one evening after threshing at the neighbors, he told Mom they would probably wrap up the threshing work at this farm the following night.

He told Mom to expect two crew members for supper tomorrow night—providing they decided to move the threshing rig to our farm when they finished late the next day.

No Big Hurry

So the following morning, Mom leisurely set about getting a head start on the meals for the next couple of days. She baked a cake, made salad dressing, checked her grocery supplies, decided what vegetables to pick

BELT POWER. It always took plenty of tractor pulley speed to power the belts that gave the threshing machine its speed and efficiency.

the next morning from the garden and took a close look at what was available in the freezer.

Even though tomorrow's noon-time meal would be her first time feeding the threshing crew on her own, Mom certainly felt ready. And confident she could handle this first-time task.

Because of our half mile proximity to the neighboring farm, Dad, my grandfather and our hired man preferred to come home at noon to eat. And to handle some important noon-time livestock chores.

"They're Coming—Now!"

So Mom was busily preparing the noon meal that day when Dad literally flew in at 20 minutes before noon with startling news:

"The men are almost finished at the other farm and you will have 12 men for dinner in 20 minutes!"

Mom later said she did not have time to panic. Instead, she had to think...and think fast.

Meat was not going to be a problem since she already had a good-sized roast in the oven—anticipating that plenty would be left over after today's dinner for tonight's evening meal for the two expected members of the threshing crew. But having only 20 minutes to prepare more potatoes and sweet corn for an additional 12 men was definitely going to be a problem.

Call For Help

She quickly called my grandmother. Miraculously, corn and potatoes were already boiling at grandma's house just down the road. She gladly agreed to immediately bring down as much food as she had prepared.

Next, Mom ran to the garden for additional cabbage, cucumbers and tomatoes. Within minutes, grandma arrived with her brimming kettles of potatoes and corn and took over preparation

"The knots in her stomach were much bigger than the lumps in the gravy..."

of the freshly picked garden foods.

Mom hastily set the table, got pickles and jelly from the fruit cellar in the basement and started a big pot of coffee. She told our family many times that she would never know how she ever got the gravy made without any lumps.

The desserts were multiple choice and no big problem. Mom's cake was not a large one. Grandma's pudding was not intended for a crowd and her apple pie was already cut for only eight hungry men. But ice cream was readily available.

Filled out with bread, cottage cheese, sliced tomatoes, cabbage salad, sliced cucumbers and green onions, the table was certainly full for the hungry threshers just before the living room clock chimed 12 noon.

She Remembered It All

Even 35 years later, Mom could still vividly recall every item of food that was quickly made ready for the hungry men on that table.

By noon, the men had arrived, washed and were sitting down at the table without standing around waiting for dinner to be served. And also with no knowledge of what those 20 minutes had done to my Mother.

After dinner was finished, she basked in the praise from all the good eating the threshing crew had enjoyed that day. She always especially remembered the praise which came from the neighboring farmer whose wife didn't bother to feed the thresh-

PICTURESQUE SCENE. While fields of shocked wheat always looked beautiful when driving down the country road, small grain harvesting certainly represented plenty of hard, hot and dusty work.

But you always had a real sense of accomplishment when the grain was in the bin and you knew you had enough feed to get your livestock through another hard Michigan winter.

ing crew that day—instead, sending them to dine at our place on just 20 minutes notice.

When the neighbor finished eating, he loudly proclaimed so everyone on the threshing crew could hear: "This was sure one heck of a good dinner. Darned if I wouldn't like to eat here when she was expecting us!"

Mom wasn't the only one who said amen to that!

Dad, my grandmother and my grandfather certainly joined in, probably along with a number of the threshing crew members who knew what had actually happened. And who had probably earlier wondered how well they would eat that day on just

20 minutes notice.

Over the years, our family has relived with great fun Mom's baptism by fire in feeding a threshing crew many times. But at the time it took place, it was hardly funny to Mom—a city-bred, city-raised bride of only a few years.

Baptism Under Fire

Yet it actually turned out to be one of her many blessings as she quickly learned to really enjoy threshing days and life in the country. After all, no meal would ever prove to be as tough to handle in the years to come.

Getting dinner was always exciting and a real challenge to

"Somehow, everyone on the threshing crew had a super meal that day on only 20 minutes notice..."

Mom. But it was never again something to dread whether preparing a meal for the four of us, 14 or 30 visitors.

On this mid-summer day—with only 20 minutes notice—Mom had proven to herself that she could handle the responsibilities of a farm wife just as well as her country friends who had been born to it.

STILL PLENTY OF GOOD FOOD. Whether it was leftover grain in the fields or freshly-planted small grains such as wheat and rye in the fall, there was always plenty of food left for wildlife, such as these Canadian geese stopping for nourishment during their annual journey to Florida in late October.

Conservation Was Always Important

WHILE CONSERVATION compliance and residue management seem to be today's agriculture "buzz words," our family has long supported the need for reducing costly erosion.

One thing is for sure...our rolling southern Michigan family farm certainly has highly erodible land.

Whether wind, soil or water erosion, generations of our family always took pains to make sure the nation's valuable resources weren't wasted—at least on our home farm.

The Marl Pit

Unless you've got some age on you, you probably won't know what I'm talking about when I tell you about the farm's marl pit. Back in the late 1940s, my dad hired a contractor with what we used to call a steam shovel to dig several year's supply of marl from a pit along the back side of our own farm lake.

For those of you who don't know much about marl, the word is rather loosely used to describe earthy or soft rock deposits which are rich in carbonate of lime.

When applied to fields, it has the same effects of ground lime-

HOME-GROWN LIME. Back in the late 1940s, a contractor dug a marl pit next to the lake. The marl was used as a substitute for limestone for liming the pastures and corn fields.

MAPLE TREES IN THE LATE 30S. Back in the mid-1920s, members of the Lessiter family dug up young maple saplings in the woods and replanted maple trees 20-yards apart along both sides of nearly 1 mile of Baldwin Road. This was long before the road was blacktopped and the wagon wheel ruts ran plenty deep.

stone in changing the pH of your soils.

I don't recall the specific dimensions of the pit, but for many years the hole looked like a large swimming pool filled with water.

After the pile of marl had dried for a few months, we used the tractor and front end loader to load marl in the manure spreader for spreading with our team of Belgian horses.

I recall the pile of marl lasted 2 or 3 years. Why we never dug out any more is a question I can't answer.

I'd guess very few farmers can talk about being able to produce their own supply of lime right on the farm, but it's a cost-cutting idea which I certainly remember as a kid.

Trees, More Trees

While I am sure earlier generations took many precautions in farming our farm's fragile soils and rolling terrain, protecting our many areas of woodlands from extensive grazing by livestock was always a key area of protection.

And it paid big dividends from generation to generation as thousands of board feet of valuable lumber were cut from the woods. Most of the barns on the farm which were built from the early 1900s on up through the 1950s came from home-grown lumber.

Pine lumber for the big barn was cut from a woods located 2 miles north of the farm.

This lumber was cut from trees growing in our own woods, were either sawed on the farm or at a nearby mill, allowed to dry and then used to construct the various barns built by the farm crews.

There were also times when trees were marked by a State of Michigan forester and then cut by contractors. This even included some very valuable logs which brought a pretty penny in later years for specialty markets.

Actually, the first conservation stories I remember my grandparents telling had to do with planting trees. They told how my grandfather and uncle Floyd transplanted maple trees 20-yards apart along nearly 1-mile of Baldwin Road back in the mid 1920s.

These trees survived many years as a valuable windbreak, provided plenty of shade and served as a natural snow fence for cars using the road. Many of these trees, especially near our farmstead, grew very tall over the years and added a pictur-

MAPLES ARE DYING. 70 years later, the maples have thinned out. Besides disease, the constant exhaust fumes from the many cars that go by each day took their death toll on the big and beautiful shade trees. The John and Donalda Lessiter house at right was built in 1938.

esque setting to the farm as you drove down Baldwin Road.

Unfortunately, many of these beautiful maple trees have now died. Yet some are still standing and looking good after having provided many benefits through nearly 70 years of growth.

Cover Crops, Rotations

Other good soil conservation practices used to reduce erosion

---

"With only four insulators, the electric lineman was told to walk across the alfalfa field..."

---

included cover crops and crop rotations.

A typical rotation for the entire farm was 1-year of corn followed by 1-year of oats, wheat or barley and 2 or 3 years of alfalfa which provided needed winter forage and spring- to-fall grazing for the dairy herd. If an alfalfa stand was still in good shape, it wasn't unusual to stretch it out over still another growing season.

Dad was always very protective of good alfalfa stands. I distinctively remember one Spring after we had sold some right-of-way to the electric company for new lines—at an exciting $10 each for the half dozen poles they placed on our fence lines out in the middle of two fields.

Several weeks after the poles had been installed, I remember Dad running out as one of the subcontractors for the Detroit Edison project started to drive his big truck into one of our alfalfa fields.

Dad stopped him and asked what he thought he was doing, driving over a very valuable forage crop. The driver said he had four insulators to place on one of the poles and that he had a right to drive in the field.

Dad told him he'd have to carry the four insulators and walk...that he wasn't going to allow the guy to destroy a valuable crop of alfalfa just because he was too lazy to walk less than 300 yards.

After a heated conversation, the driver parked his rig, picked up the insulators and stomped off through the alfalfa for his walk.

Keep It Covered

Our cover crop program was also important in providing winter protection of the soil. We normally seeded wheat in late September into fields where corn had been cut earlier that

WHEEL TRACK PLANTING. While this photo was not taken at Lohill Farm, it shows the wheel track planting system used by Dad for several years back in the early 1950s.

After moldboard plowing, the next trip across the corn field was made with our 2-row planter without any of the usual disking, harrowing and cultipacking trips—cutting out at least three moisture-robbing, expensive tillage trips. It was always a rough bumpy ride, but this reduced tillage idea really worked and was the forerunner of today's minimum till and no-till systems.

month for silage. The fields were tilled and seeded in late September before we started cutting corn in mid-October for grain.

Alfalfa would be overseeded the following spring in early April into the wheat. Or we might seed oats or barley along with an alfalfa seeding.

Wheel Track Planting

In the early 1950s, Michigan State University soil scientist Ray Cook came up with the idea of wheel track planting, the forerunner of today's minimum tillage and no-tillage programs. It amounted to plowing a field, then making the very next trip over the field with a 2-row corn planter!

Cook's wheel track idea suffered from a few problems—including being plenty tough on your body from all of the bumping, damaging machinery and often giving poor seed placement. Yet moisture conservation was definitely a big benefit.

Even back in the 1950s, the idea proved that many trips with the disk, harrow and cultipacker weren't really necessary.

Always a pioneer in trying new farming ideas, Dad decided to give wheel track planting a try. We tried it on a number of fields for 2 years, then gave it up mainly because of the rough ride and erratic seed placement.

But this keen interest in reduc-

ALFALFA SEEDING CLOTHES. Dad always liked to look good for photos and even put on a tie the day a daily newspaper photographer came to take this shot of alfalfa seeding.

ing costly field trips in the early 1950s paid off in later years when no-till was used on a number of the farm's corn fields.

All Tied Up

Dad's tradition of wearing a tie brings me to another soil conservation tale that took place on our farm one April afternoon.

Dad always liked to get dressed up when he went to town, to a meeting or to visit practically anyone—something he still does today.

One Spring day, officials of the Oakland County Soil Conservation and Water District had arranged for a photographer from the Pontiac Press daily newspaper to get a photo of Dad seeding alfalfa. Even today, many years later, I have to laugh when I pull out a copy of the newspaper clipping.

There sits Dad on the tractor pulling the grain drill through the field—wearing a pretty flower-decorated tie. He's the only farmer I've ever seen driving a tractor and working in the field while wearing a tie!

One More Tale

Before I wind up this story on conservation practices used at Lohill farm, there's one more story I need to tell. While it happened years ago, it has an impact on some of today's modern-day farming practices.

Toward the end of the farm lane, we had a 10-acre muck field. It was plenty wet and was used in earlier years for grazing the farm's beef, sheep and dairy cattle.

When I was a kid, Dad decided to drain off this muck ground so we could plant high-value crops such as potatoes on

> *"Everyone always gets a good laugh when they see Dad seeding alfalfa while wearing a neat shirt and tie..."*

MUCK LAND OR WETLANDS? This 10-acre field of muck served many purposes in the 100-year plus history of Lohill Farms. But a plugged drain tile changed the way in which federal government officials would view this field.

stayed at school one year during Spring vacation to take a week-long special course dealing with farm dynamiting.

Based on what he had learned in that class, he dynamited a drainage ditch along the western edge of the muck field which flowed into the old tile system.

Without practically any shov-

this rather fertile soil.

At the south end of the muck field, an elaborate tile drainage system had been installed years earlier which drained into the lake a quarter mile away.

When Dad studied agricultural engineering at what was then called Michigan Agricultural College in East Lansing, he

10-ACRES OF WATER. It was a shock to pull over the hill and see a miniature lake where you instead expected to see the muck field you helped farm as a kid.

NOW A WETLANDS. Because of the plugged drain tile, this field would be designated as a wetlands area and changes could not be made without the permission of federal government officials. All simply because of a plugged tile line.

eling, we ended up with a 500-yard drainage ditch that drained water off this field and let us plant a high yielding potato crop in this field.

Gardening Disaster!

One year, Dad even got the idea of planting the family garden down here, but it turned out to be an absolute disaster.

First off, the area had to be fenced to keep livestock and wild animals from feasting on all the vegetable crops.

Second, many vegetables went to waste because it was so far from the house that Mom or Dad had to get in the truck or use the tractor to visit the garden.

Third, it was the weediest garden we ever had because nobody ever was back there to weed it.

During the day, the farm crew

PROTECTED WETLANDS. If someone today simply unplugged the tile lines and drained this pond into the farm's bigger lake, they would be risking a huge fine and even the possibility of jail time.

A BEAUTIFUL VIEW. The old tree silhouetted at sunset in the 10-acre pond is certainly a beautiful site. But it's a scene that simply wasn't there when many of the Lessiter generations were growing up on the farm.

was too busy and after supper, it was too far to go to handle the weeding chores.

Having a garden one-half mile from the house was an idea Mom put up with for only 1 year.

Changing Times

With Dad pretty much retired and my sister and her family's beef cattle roaming most of the farm, things changed on the old farmstead. I saw one of the big changes firsthand several years ago when Dad and I drove back to the muck because he had something interesting he wanted to show me.

I was shocked when we pulled over the hill and I saw what used to be the old muck field. About half of the 10-acres were completely covered with water!

Dad explained how the old tile lines apparently became plugged and the field was no longer draining into the lake. If someone wanted a beautiful pond with ducks and geese, they had it now. And they might have it for years to come whether they really wanted it or not.

The reason is because of the current wetlands issues being fought by environmentalists and in Congress. During a later trip to Washington, I told top officials of the Soil Conservation Service about this old muck land and what had happened to it.

They informed me it was now a wetlands and couldn't be turned back into profitable cropland without permission from the Soil Conservation Service, probably the Army Corps of Engineers and who knows what other government agencies.

I protested that all anyone had to do was clean out the tile lines to get the water flowing once again off the muck field. They told me it wasn't that easy since the muck land now qualified as a protected wetlands area. And that the Lessiter family could go to jail if we cleared the tile lines without first getting governmental agency permission.

So if you've got any fields like this where blockage in an old drainage system may turn it into a wetlands area, be leary of what is happening. You might want to get those old tile lines fixed before you find part of your farmland with a brand new wetlands classification.

Mechanical Skills Were Critical

EVEN WITH THE change from horsepower to tractor power, the Lohill Farm operation always required plenty of machinery and equipment to get the cropping and livestock work done in an effective, timely and profit-building manner.

From harnesses to whippletrees to reins with the horses, to tillage equipment to shop tools to seeding equipment to feed processing, there was always lots of equipment and machinery to keep in tip-top operating condition.

Generation after generation, it was critical that the Lessiter family members develop the necessary mechanical skills to keep everything purring. If you had to head for town each time there was a problem, you couldn't expect to make much profit from farming.

Being independent and talented enough to make farm ma-

DAYS GONE BY. Real horsepower in the form of teams of Belgian horses were the main source of power on the Lessiter family farm for many years. Harnessing and shoeing the teams was a critical aspect of farm life until tractors eventually took over all of the farm work.

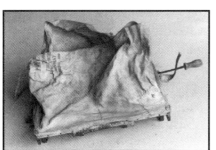
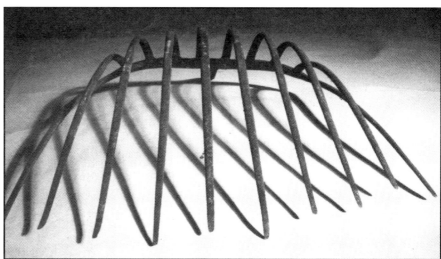

CLOCKWISE FROM TOP LEFT. Feed grinder, farm shop boxes and fence fixing box, corn silage fork, bolt for plowshare, replacement moldboard and plowshare, hand-carried, hand-powered alfalfa seeder.

chinery repairs was an important part of the family's farming success. Usually, there wasn't much that needed doing on the farm that some family member couldn't tackle successfully.

Old Tractors Still Going Strong

HAVING BEEN A teenager at Lohill Farm in the early 1950s, I fondly remember the slick new Ford Jubilee and Farmall Super H tractors coming off the delivery trucks.

I don't remember all of the economic details, but 1953 and 1954 must have been extremely good farming years for us because we bought new tractors both years. In 1954, the decision was made to switch to baling hay, and this meant buying the Farmall Super H and a pto-powered International baler.

Both tractors are still going strong today at my sister and brother-in-law's farm west of Lapeer, Michigan. As you might guess, both tractors would bring more dollars today than Dad paid for them 45 years ago!

Ford Farming For Years

Besides a couple teams of Belgian horses, Dad and Grandpa Frank had farmed with several older Ford and Fordson tractors.

In fact, the Lessiter family's relationship with the company goes back to 1902 when Henry Ford visited the farm to examine a Shorthorn bull as a potential herd sire for his herd. That was a year before the Ford Motor Company was officially formed

FARM POWER. A Ford Golden Jubilee and an International Harvester Super H were mainstays of the farming operation in the early 1950s. Both are still going strong today.

FORD JUBILEE. This newly-stylized and upgraded tractor was introduced to celebrate the 50th anniversary of the Ford Motor Company in 1953.

and four years before ex-farm boy Henry Ford started thinking about producing tractors.

By 1917, Ford was building the Fordson model F tractor in Ireland. It had an 18-horsepower, four-cylinder engine and featured a three-speed, worm-reduction gear system. In 1918, 6,000 of these tractors were built at a Detroit factory.

Within a few years, Ford grabbed 70 percent of the world's tractor market and produced 101,898 tractors in 1923. The 1922 Fordson sold for only $395!

The 50-Year Model

When the 1953 Ford Jubilee tractor was introduced to celebrate the 50th anniversary of the Ford Motor Company, Dad was among the early buyers.

Larger and heavier than the 8N tractor it replaced, the Jubilee had an overhead-valve, four-cylinder, 134-cubic-inch engine. Completely restyled, it no longer resembled the popular 8N Ford tractor or the Ferguson TO-30, a major competitor which came out in 1951.

What I remember most about this tractor was the "Golden Jubilee Model 1903-1953" nose medallion used to promote the founding of the company in 1903.

This medallion was made only during the 50th anniversary year and there was a revised star-encircled medallion on the 1954 tractors—about the only change made in that year's production.

A non-live pto was standard on the Jubilee tractor and a live pto was optional. The brakes were improved and a better governor was built. The muffler was moved from under the engine to under the hood to reduce the chance for a hot muffler causing fires in dry straw. The tractor also featured a Ford-designed, camshaft-driven pump mounted on the engine.

These Jubilee tractors were painted red and gray, an exciting change from the gray Fordson and Ford tractors seen for years. So much for the auto philosophy

"Red and gray was an exciting change from the gray Fordson colors seen for many years..."

of Henry Ford Sr., whose credo was, "Any color as long as it's black."

Built From Necessity

It was more than just coincidence that a new Ford tractor made its debut during the 50th anniversary year of the company. To fully understand the situation, you need a little history.

The revolutionary three-point implement control system with automatic compensation for changing draft loads with implements was invented by Irishman Harry Ferguson. He had partnered early with Englishman David Brown to build the Ferguson-Brown tractor.

When the partnership didn't work out, Ferguson sold Henry Ford on the idea and they quickly struck a gentleman's agreement to build a new tractor incorporating the Ferguson hydraulics.

SUPER H. This International Harvester tractor was only manufactured for three years. The narrow tires were the style long before bigger, wider tractor tires came into play.

The first joint-venture 9N tractor rolled off the assembly line in June of 1939 and more than 10,000 tractors were sold that year. The 9 meant 1939 and N was the Ford designation for the tractor. To hold the price to around $600, parts from Ford car and truck assembly lines were used on the tractors. The tractor proved very popular, but World War II material shortages and rationing meant low tractor inventories.

The designation was changed from 9N to 2N (for 1942) to reflect simplified changes and to get around the War Production Board's price controls.

Following the war, Henry

> *"A legal battle between Henry Ford and Harry Fordson led to development of new tractors..."*

Ford Sr. was forced out and his grandson took over the company controls. Faced with nose-bleed-sized losses, the decision was made in 1946 to disband the tractor partnership with Harry Ferguson. The younger Ford soon figured out Ferguson was the only one making any money in the deal.

Ford had built tractors at a fixed price for Ferguson, who then sold tractors and equipment to farmers through his Ferguson dealers, one of which was located in Pontiac, Michigan.

As a young kid, I remember sitting with Dad in the Ferguson tractor dealership—actually the garage behind the dealer's house in downtown Pontiac.

The Ford goal was to build a completely new tractor to go head-on with whatever tractors Ferguson could piece together. By 1948, Ford offered a line of implements under the Dearborn Motors Company name.

I remember seeing farm magazine ads for Dearborn Motors farm machinery. Since the city of Dearborn was the home of Ford Motor Company, it wasn't hard to figure out this tie-in with the new farm machinery arm of the Ford organization.

By then, Ferguson had developed the Ferguson TE-20 for Europe and soon followed with a plant in Detroit to build the TO-20 American version.

$251 Million In Damages

In late 1946, Harry Ferguson sued Ford for $251 million. Because Ford would no longer provide tractors, he felt he earned

40 PLUS YEARS OF HARD WORK. By today's farming standards, the International Harvester Super H certainly wasn't any high horsepower tractor. But it did a tremendous amount of work over the years, played a key role in changing the farm's haying system and is still going strong as a "chore tractor" more than 45 years later.

these dollars for loss of business and patent infringement.

Hydraulics Were The Key

Yet the real bone of contention was the hydraulically controlled three-point system, the only point on which Ferguson eventually won in court. This led to a court settlement of $9.25 million, much less than what Ford paid lawyers on the case.

Most importantly, Ford was required to come up with a new hydraulic system by 1953. This led to the decision to have Ford engineers design a better hydraulic system and to launch the new Jubilee tractor in 1953.

Even as a teenager, I could tell that the tractor featured a pretty darned close copy of the Ferguson three-point hydraulic system. But it somehow withstood later court deliberations.

The Farmall Super H

When we switched to baling hay in 1954, the Farmall Super H tractor was needed for pto horsepower. Plus the pto shaft of the Ford Jubilee tractor simply wasn't high enough off the ground to handle the baler's power needs.

With higher engine displacement and more power than previous H model tractors, the Super H wasn't manufactured for very long by International Harvester. It was introduced in 1952 and production was halted in 1954.

These H series tractors were advertised by International Harvester as a "two-plow tractor in any field." They were probably International's best-selling full-size tractor with more than 350,000 sold from 1939 to 1954.

The completely revised 1939 tractor design better served the changing needs of the American farmer. Features included increased horsepower, more reli-

"By 1954, the letter line was discontinued and the company would never be the same..."

ability and utility. New manufacturing materials provided longer life and many engine improvements were made.

The H tractor was a 1939 replacement for the much-loved International F-20 tractor. In 1940, you could buy a model H tractor for $962 with rubber on all four wheels. This cost $172.50 more than tractors that came with four steel wheels—something many farmers had to settle for during the early 1940s when rubber was in short supply during World War II.

All Over By 1954

By 1954, the "letter line" was discontinued and replaced by the "hundred" series of bright red tractors. Yet there's much more to International's role in the tractor field at that time than just changing the way tractor models were named.

In hindsight, unfortunate business moves by International in the mid-1950s probably brought an end to the company's dominance of the U.S. tractor market.

Branching out into construction equipment, home appliances and other products meant the company was short of much needed investment and manufacturing funds.

This soon led to the company falling behind the U.S. market in the one area which made the company great—production of those bright red tractors.

Driving The Old Oakland Cars

THE FIRST CAR the Lessiter family ever owned was bought in 1911. It was a four-cylinder Oakland car, like the one shown here. And it was the first admission by the family that perhaps horses wouldn't be the way to travel forever.

It had white rubber tires and rode pretty hard over what passed for country roads in those days. The name came from the location of the manufacturing facility in Oakland County.

THE 1911 OAKLAND CAR. Manufactured just 10 miles south of the farm in Pontiac, this 1911 Oakland was the first auto ever owned by the Lessiter family. When the car needed service, family members drove it back to the plant in Pontiac. The Oakland later became part of the Pontiac Division of General Motors.

RACING SOLD CARS. In the early days of the 20th century, auto manufacturers sold the merits of their cars with local and crosscountry races. A group of folks turned out on a Sunday afternoon to watch this 1911 Oakland Roadster, similar to one purchased by the family, show what it could do out on the road.

The Oakland was actually the forerunner of today's Pontiac automobile. These Pontiac, Michigan, carmakers sold the company later on to General Motors.

If something went wrong with the car, Grandpa Frank would simply drive it 10 miles south to the factory and have it repaired. Actually, the so-called "factory" was only a few old run-down farm-style sheds which were located at the corner of Oakland and Baldwin Roads.

Since it was a summer touring car, it was put up on blocks during the winter months. That meant going back to driving a horse and buggy during cold, snowy weather.

The "Sensible Six"

The family's second car was the "Oakland Sensible Six," another touring car.

A third car was bought in 1922. It was an Oakland sedan—a car that could be driven all winter long. About that time, the family finally gave up using horsepower and a buggy or sleigh for winter driving.

The Chevy Roadster

By the time Dad was a senior at Oxford High School in 1925, he had his own Chevrolet Roadster to drive back and forth. It was an 8-mile trip and took about 30 minutes.

By this time, Grandmother Wiser had moved from the Seymour Lake farm to a house in Oxford. Dad had lived with her during his first 3 years of high

---

The Fordson Sale!

Back in the mid 1940s, Dad bought a new tractor and put the well-used Fordson up for sale. The price was $300.

When a farmer came to buy the tractor, he pulled out a $500 bill and asked Dad if he had change.

When Dad said no, the farmer put the $500 bill back in his wallet and then pulled out three $100 bills to complete the deal.

As a young farm kid, that left quite an impression on me. I'd never seen any farmer carrying that much money in his wallet!

SAME GUY, SAME TRACTOR. As a teenager, Frank Lessiter spent many hours driving this tractor. Both are still going strong 40 plus years later.

school, but decided to commute during his senior year.

Grandma Norah was always sending food to Grandma Wiser. So Dad would leave school, drive down to her house at noon, deliver the groceries and eat a

FORD TRACTORS, CAR, TRUCK. For a few years during the 1950s, every vehicle on the farm was a Ford. Besides the Ford car and truck, there was this 1953 Ford Jubilee tractor which marked the 50th anniversary of the Ford Motor Co. It is still going strong today, more than 40 years later.

quick but specially prepared lunch with her.

Even in those days, Grandma Wiser was 80 years old and still living alone.

More New Cars

I remember when we were a real Ford family in the late 1940s. We had continued to drive a 1941 Chevrolet car until 1948 when Ford came out with the first post-war models. We purchased the very first one sold by Millman Ford in Lake Orion that year, which turned out to be a big mistake. That car was truly a "lemon!"

There was always something wrong with that new car. It seemed it was in the repair shop almost as much as we were able to drive it.

Only a year later, Dad traded for a 1949 Ford, figuring they'd got the bugs worked out of the cars by this time.

A few years later, we traded the old GMC pickup truck for a new Ford pickup. The Harry Ferguson tractor was traded for a 1953 Ford Jubilee tractor which is still going strong on my sister's farm.

For a few years in the early 1950s, every vehicle on the farm was a Ford.

Hired Hands Essential For Farming Success

GROWING UP at Lohill Farm, I remember two of the hired men who helped operate the family farming operation.

Later during high school when times got tough in farming, I even took on the role of the hired man for a year before heading off to college.

After going to high school and working full-time on the farm at the same time, you can see why going off to college was like a vacation for me. Like most farm kids, I'd never had so much free time in my entire life!

Stairway To The Stars

While I was too young to remember his name, the hired man who had the most impact on my life worked on the farm when I was about two years old.

On a cold, blustery January day, the furnace had gone out at home and Dad and the hired man were down in the basement trying to figure out how to get it going again.

I was riding my rubber-tired tricycle in the kitchen, got too close to the top of the basement stairs and tumbled down the stairs toward the cold, hard con-

THE TENANT HOUSE. For years, the farm's hired man and his family were provided with free housing in this structure which was the original 1854 home on the Lessiter Centennial Farm.

crete basement floor.

The hired man heard me bouncing off the stairs on my trike, ran over and caught me just before my head hit the concrete. I was banged up a little, but nothing serious.

It could have been much worse if the hired man hadn't been so quick to react. As a result, he probably had as big an impact on my life as any hired man who ever worked on the farm.

Later, several other hired men lived with their families in the old tenant house, the original house on the family farm. Since

> *"It was a sad day for everyone when Dad was forced to say goodbye to the hired man..."*

they were there during my younger days, I don't recall all of them, although one or two used to come back occasionally and visit with Dad.

The Long-Termers

About the time I started school, the second hired hand and his family that I remember so well came to the farm. They stayed the longest—up until the mid 1950s. About that time, dairying got real tough with low milk prices and high costs and Dad wasn't making any money.

Finally, Dad had to cut back and let this employee go, although he continued to rent the tenant house for some time before they purchased a house in Lake Orion.

This hired man usually didn't have much to say, unless you got him mad, which I managed to do

RING AROUND THE TREE. This is the tree in front of the barns that the hired man danced around while being chased by a huge Holstein bull.

a time or two. It seemed like he always wore a Stroh's beer shirt and it soon became his idenity.

A neighboring farmer liked our hired man, but also liked to tease him. One time our hired man got so mad at the teasing that he took off his hat, jumped on it and swore up a storm. And the madder he got, the more he was teased.

The Bull And The Tree

One time a bull which the

hired man was leading got loose in the farm yard and he high-tailed it for the tree where the rope swing hung for years. The bull was right behind him.

The bull would go around the tree and start to charge and the hired man would move to the other side of the tree. Dad says it looked funny, because the worker was scared and moving as fast as he could.

Someone finally distracted the bull long enough so the hired man could run for home.

"I can see him yet," recalls Dad. "He jumped over the top of that garden gate in the fence. If he'd had to jump it under normal conditions, he would have never made it. But he was so scared of that bull and wanted to get to safety, that the gate wasn't going to stand in his way."

Laundry Basket Concerns

My Mother was sick for several months when I was seven years old. During that time, the hired man's wife did most of our

THE OLD FARM HOUSE. Built in 1854, this house was the original family home at Lohill Farm. It underwent a number of additions and remodelings over the years and served as home for several generations of Lessiter family members through the 1990s.

family's laundry.

This lady was pretty practical and I guess she knew what I needed in the way of new clothes. I was surprised when I got three pair of underwear as a Christmas present from their daughter that year!

Playing Telephone Games

For several years after the hired man's family moved into the tenant house, they didn't have a telephone. The hired man's wife would come to our house and use ours whenever she needed to make an essential phone call. Mom used to say some of those calls got pretty long-winded.

Later, they became one of the dozen or so families who shared a party line. Sometimes you would hear a click after our telephone rang and you knew somebody was listening to our conversations. Mom never said much about it, but she'd later hear things that had only been discussed with someone else in a telephone conversation.

Shocking Experience

There was also the time the hired hand's wife was climbing one of the fences and discovered quite by accident that it was an electric fence. She'd get a shock from the electric fence which would make her jump up in the air and then she'd come back down on the electric wire.

While it was funny to see, it wasn't something you would wish on anyone. After a couple of shocking jumps, she somehow got off the fence.

Chickens And Skunks

The story I remember best about the hired man dealt with the chicken range shelter we placed in the pasture field next to our house. After the young chicks had some age on them, we would put them in a range shelter out in one of the fields so they could run outside while developing for their future career as laying hens. Besides being fed and watered, they had the chance to soak up the sunshine and nutrients found in the grass.

This particular time we had the range shelter in the same field used for pasturing the dairy herd. We had run an electric fence wire 12 feet behind the range shelter to keep the cows from scaring the chickens and eating their feed.

Dad and I came running when the hired man told us there was a skunk under the range house. Since skunks loved to dine on baby chicks, we wanted to get rid of the problem.

Dealing with a skunk is no easy task. You definitely want to trap it or get rid of it, but you

---

"Dad replaced the hired man with me for only $8 per week..."

---

don't want to encourage the animal to leave that long-lingering, terrible smell behind either.

As you might guess, that's what happened. The three of us were trying to chase the skunk away when it suddenly raised its tail and warned us it was ready to let go with that dreaded odor!

When the hired man saw the skunk's tail go up, he backed up—right into the electric fence! The shock from the fence caused him to jump forward five feet and that's when the skunk hit him square in the face with that bad odor.

Dad and I burst out laughing which only made him madder. But we couldn't help it!

Yet it was no laughing matter because he got hit full blast with the skunk odor. His eyes were running and he was in pain.

Finally, he washed his face and the smell started to go away. But he stayed mad at Dad and me for some time for laughing about the whole situation.

School Work, Farm Work

He was a good worker and the dozen or so years he spent on the farm were good ones for everyone. It was sad when the time came to cut back on hired labor because of the bad dairy situation.

Dad's idea of a low-cost replacement was me—a teenager who would soon be starting his senior year of high school.

Actually, it worked out quite well and helped me learn more about hard work and responsibility. I continued to get up in the morning and help Dad with the chores.

My school day was arranged so I was finished with classes at 1 p.m. This meant I could come home and work on the farm all afternoon before doing my school work that night.

While it made for some pretty long days, Dad jumped my allowance from $2 to $10 per week. This was the biggest allowance any member of my senior class enjoyed back in 1957.

Of course, that was a pretty good investment for the family, too, considering Dad had replaced a full-time hired man for only $8 per week.

His Name Was Leon McGrath

A FEW WEEKS BACK, I drove back home to Michigan to see Dad. During my visit, I learned that he was having troubles with the water system at the house.

The situation reminded me of a name from the past—Leon McGrath.

He Could Do It All

Whenever we had a water problem on the farm while I was growing up, we would call Leon McGrath. This highly talented handyman lived 4 miles south of the farm on Baldwin Road and did occasional problem-solving work for us for more than 30 years.

McGrath was among a vanishing breed of handymen who made life much easier for many farmers in earlier years.

He was always ready to tackle any job, would never quit until he had it figured out and was pretty dependable, especially when you were in a real jam with cows to be milked or calves desperately needing water.

Leon would soon show up in dirty old bib overalls with a couple of days of whiskers growing on his face.

He would climb down out of his battered old truck which seemed to hold any tool he would ever need to fix just about anything. (The tools were always piled in such a heap on the

MASONRY SPECIALIST. Leon McGrath could do everything from pouring cement floors, to setting concrete block, to repairing a stone wall like this one that served as the big barn's foundation.

177

HIT WATER AT 12-FEET. When a well was dug in the 1970s by hand near the original house on the farm, they struck water at a very shallow depth. Yet Dad remembers when the family home was built in 1938 just 100 yards to the north. The well men went so deep in search of water that they had to finally give up and move to a new drilling location about 50 yards away.

truckbed floor that nobody but McGrath could ever find what he needed).

Then he'd close the truck door, roll a plug of tobacco around in his mouth, spit some tobacco juice on the ground and say to Dad, "What's the problem?"

Soon, the water system would

"Leon helped build the new house back in 1938 and worked for a wage of only 80 cents an hour..."
---

be going full tilt and the cows would be back up at the tank enjoying a cold drink. It didn't seem to matter what the problem was, McGrath would soon have it fixed and running.

As a kid, I can remember McGrath spending occasional days at our farm:

★ If we had concrete to pour, Leon could put the proper slope on a barn floor with the best of them.

★ If we had concrete block to be laid, Leon would always end up with the straightest wall possible.

★ If there was carpentry work to be done, Leon always swung a strong and steady hammer.

★ If there was plumbing work to be done, Leon knew a pipe wrench from a pipe cutter.

★ If we were having difficulty with a ventilation fan or electric motor, Leon would soon have it running as if he had built it himself.

★ If we were remodeling the barn or designing a new way of handling cattle or installing new milking parlor equipment, Leon, as dependable as the sunrise, would always be there.

★ If the water pipes were frozen on below zero winter days, Leon soon had them flowing so the cattle could drink, just as if it

LOVED TO DO BARN WORK. The very best at barn building and repair work was the most accurate way to describe the many talents of Leon McGrath. He had a special touch when it came to tackling these kinds of projects and he did extensive barn work at Lohill Farm over the years.

was a hot, summer day.

"He was really handy and he did a lot of work for us over the years," Dad told me as we were waiting a few weeks back for the well man to get the water system running again at his house. "There really wasn't much he couldn't do."

80 Cents An Hour?

When Dad and Mom built the family's new house in 1938, McGrath helped with the construction for several months and was paid 80 cents per hour, a good wage in those days.

When we built the new barn in the late 1940s and the trusses came in a railroad box car, I remember McGrath was there to help unload, bolt and set the trusses in place.

While he would tackle any kind of farm job, he loved the chance to get involved with new construction projects, such as the heifer and bull barn we put up in the late 1940s.

"When the posts in the potato cellar basement were starting to droop, Leon jacked them all up, poured concrete around the base,

"Regardless of the work that needed to be done, Leon could always find the solution..."

reset all of the the posts and the barn is still in good shape today," says Dad.

Most people around town thought McGrath was a character, but he bailed us out of plenty of equipment jams on the family farm over the years. It could get scary when there was a problem which needed immediate attention and nobody could locate him. But after a few telephone calls, McGrath would always show up.

He was always ready to tackle any kind of project, was seldom in a hurry to move on to the next job or head for home at the end of the day, and could fix just about anything that needed work on the farm.

"He was never too particular about himself or the way he dressed, always seemed to need a shave and didn't get along all

PLENTY OF REPAIRS. Having been built as far back as the early 1900s, some of the old barns often needed plenty of reparir work. Leon McGrath's remodeling tasks ranged from shoring up drooping posts, putting a new shingled roof on one of the barns or tearing down a building that had finally outlasted its usefulness on the family farm.

that well with his wife," recalls Dad.

He Liked Our Family

"But I think he always enjoyed coming to our place and helping us solve problems. He always enjoyed staying for lunch since your Mom always cooked up a good meal for him. I wouldn't be surprised if that wasn't one of the reasons he always helped us out so much."

"And it didn't matter what the problems were—Leon could always somehow get them corrected. There wasn't much he couldn't do. And you simply can't find these kind of valuable people today."

The Oldest Farm In America?

While the Lessiter family farm got its start back in 1853, it doesn't even come close to being the oldest farm in the United States.

Thart honor goes to the Tuttle family farm in New Hampshire which dates back to 1632. Since that time, eleven generations of the family have worked the land.

The Tuttle farm's more than 360 years of ownership by the same family dates back almost to the beginning of Colonial times. And its history has closely paralleled the growth of the nations' farming industry.

The family's start with the farm occurred back in Bristol, England, when King Charles I granted an apprentice barrel maker—John Tuttle or "Immigrant John" as he's now called by family members—the right to about 20 acres on "Hilton's Point."

This was a 9-year-old settlement located between the Cochero and Bellamy Rivers in New Hampshire. John emigrated on the ship Angle Gabriel and the rest is history at the farm which is still being farmed by the family 60 miles northwest of Boston.

Cold Concrete, Late Friday Nights

IT'S BEEN A LONG, long time. But I still remember those good times with some of my friends that often followed Friday night football and basketball games when I was a high school junior and senior.

Mom and Dad never said much if I came home late on those exciting Friday nights. I'm sure they worried as most parents and grandparents do (now I know what that's all about) as to what their son or daughter was doing, where they actually were and what time they might finally find their way home. But the two of them never really said much to me about it.

90 Minutes Of Grace

If I was getting home after midnight, my goal was to be in bed no later than 1:45 a.m.

Mom had a cuckoo clock in the living room and it chimed once at 12:30, 1:00 and 1:30 a.m. So If she heard me come in and the cuckoo chimed only once, she really didn't know what time it was. Yet I don't think she was so naive to think it was always 12:30 a.m.

When I'd stayed out too late on a Friday night, Dad had a simple way of dealing with it.

For some reason, it took me a long time to really figure it out.

COLD WINTER LOOK. This was the farm's gravel pit where family members shoveled tons and tons of gravel to mix with cement for various farm construction projects over nine decades.

SATURDAY TREAT. Staying out too late on a Friday night as a high school student usually meant we'd be pouring concrete all day Saturday. By the time high school graduation rolled around, much of the barnyard was concreted.

And then I had to decide whether staying out late on Friday night was really worth it.

Every Saturday, Dad would rustle me out of bed by 6 a.m. to help with milking and feeding the dairy cows and calves. About 7 a.m., we'd find our way into the house for breakfast. As we chowed down on a hot, tasty meal, Mom would ask what the two of us would be doing today.

Those Terrible Words

If Dad told her we were going to haul rock and pour concrete, then I knew I'd stayed out too late the night before.

From the time we got the morning chores done about 8:30 a.m. until noon, we'd be loading big, heavy rocks onto the wagon from one of the huge piles found along the fence rows on one of the fields in the back of the farm, haul them up to the barnyard and lug them around until we had them perfectly set where we were going to pour concrete in the afternoon.

By the time we went to the house for the noon meal, I'd be wondering why I'd stayed out so late the night before. While Dad was usually the one who could lie down every day and sleep for 30 minutes after dinner, I joined him for a few minutes of much needed rest on these Saturdays.

By 1 p.m., we'd be back pulling the old cement mixer out of the barn and gearing up to pour concrete. And that didn't mean calling the ready-mix truck.

No! We had our own small cement mixer, although Dad always reminded me of earlier days when they always mixed concrete shovel after shovel in a big steel tub. In fact, I've done that a few times back on the farm with small patching jobs when we didn't bother to get out the mixer.

The next order of Saturday afternoon business was to take a wagon down to the farm's small gravel pit and shovel on a big load of gravel before heading back to the barnyard.

Muscle Aching Recipe

Next came an afternoon that meant following an exhausting ritual: toss in a little water from a bucket, add a shovel of cement and six shovels of gravel. Then repeat the process over and over.

The only rest came when a batch of concrete was mixing. Then I would dump the fresh concrete into a big wheelbarrow and push it over to where Dad was working and floating the concrete. The wheelbarrow was so heavy I could hardly lift it.

I still shiver when I recall my sore, cold and chapped hands covered with wet concrete as I helped finish off the latest pour. Just because the temperature was below freezing didn't mean you couldn't pour concrete!

While we started milking by 5 p.m. on most days, it always seemed on these cold, dark and often rainy fall and winter Saturday afternoons we were still floating concrete at that hour.

This meant it was usually 6:30 or 7 p.m. before we had the cows milked and the two of us finally headed to the house for a well-earned supper and the welcome end of another hard day.

Saturday Night Fever?

About this time, my school friends would be telephoning to figure out what we were going to do on Saturday night.

After having stayed out late on Friday night, hauling rocks plus gravel and pouring concrete all day Saturday along with the normal dairying chores, the answer was easy.

I wasn't ready to do anything around town tonight—I was going right to bed!

Winter Was Never Easy

FOR MANY GENERATIONS, the Lessiter family battled the weather while keeping farm work on schedule and a wide variety of livestock fed and bedded down during the long winter.

While winter can be a time of beauty on the farm with remarkable Jack Frost ice and snow renderings dotting the countryside, these same weather conditions can bring some of the most troublesome times of the year.

Tough Conditions

Extremely cold temperatures often meant frozen water pipes needed to be thawed, livestock had to be moved inside and ice in water tanks needed chopping. Newborn pigs, lambs and calves needed to be protected, week-old chicks had to be kept extra warm and the kids neded to be bundled up to avoid frostbite when waiting for the school bus or doing chores.

Drifting snow could mean farm families were snowbound for days—as was the case in 1947 when a stranger spent three days in our home. There was no way to go to town to grind feed, get milk to the creamery, buy food from the store or check out the latest newspapers or mail.

Ice storms were even more troublesome. Downed electric and telephone lines meant no electricity to run the water pump or milking machines, to heat the brooder house, warm the house or prepare meals. And there would be no way to call the veterinarian when the very best cow in the herd needed help in delivering a backward calf.

Yet one of the advantages of farming in a fairly suburban area like ours was that the electric

A TIME OF BEAUTY. Jack Frost could certainly paint a pretty portrait on the farm, but winter also meant plenty of hard work constantly battling the weather.

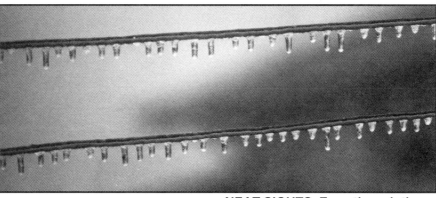

NEAT SIGHTS. Even though the winter weather could create a masterpiece of art, everyone on the farm always looked forward to spring and summer.

lines were quickly back in place after an ice storm.

While it's upsetting to a farmer with cows to milk to be without electricity, Detroit Edison Company officials knew it was much worse to have as many as 500 home owners screaming that they had no heat or light in the house.

Because of our suburban location, we normally did much better in getting electricity restored than farmers in more remote areas when they were hit with ice storms that brought downed power lines.

Winters Were Worse

Even though we had winter weather concerns, it was even worse for previous generations. That's why many farm folks don't think today's winters are as severe as they used to be.

"Winter was a big pain to the family back in the first half of the century," says Dad. "We'd let the cows out in the morning and again at night to drink. The water in the stock tank would freeze up and we'd spend an hour or two nearly every day chopping ice so the animals could drink."

In those days, there was a large stock tank over by the hen house. Yet the water for this tank underwent an extensive trip before reaching this stock tank.

The water was pumped by an old, wooden windmill located over a well alongside the old milkhouse in the center of the farmyard. While this well was only 30 yards from the water tank in the barnyard, the water traveled 10 times that distance.

Water was pumped from the windmill across the road to a big tank sitting on one end of the enclosed back porch at the big house. The overflow would be piped back across the road and down the hill to the stock watering tank.

"Your grandfather later put up a taller steel windmill to replace the wooden one," says Dad. "It could pump more water because the windmill was higher and would catch more wind currents.

"Later, they went to an electric pump at the well and a tank in the basement of the big house, but water still made the long trip across the road and back.

"One winter, the pipes froze up under the road. We rigged a temporary system to get water to the stock tank and had pipes sitting on top of a bunch of sawhorses. We had to keep the windmill pumping all the time and the water flowing constantly to keep it from freezing in those above-ground pipes."

Later, a concrete stock tank was built with a wooden canopy next to the big barn. Ice seldom formed in this partially insulated tank. Maybe it helped that Grandma Norah's goldfish spent the winter in this tank and kept the water moving.

When Harvesting Actually Came In Mid-Winter

AS DAD remembers, three or four men, a team of horses, a couple of long-bladed handsaws, pike poles, spuds, chutes, tongs and a sleigh were needed to handle mid-winter harvesting on the home farm.

If you haven't guessed by now, the cold harvesting of this mid-winter "crop" which our farm family always did in January and February was ice. And our family didn't have far to go in the old days to cut ice since there were two lakes located on the farm.

Like most crops, weather played an important role in producing this crop. Unlike the hot summer weather needed to mature the corn and alfalfa grown on our farm, we really needed plenty of cold, winter weather.

Dad says the family cut ice every winter off Dennis Lake up into the mid 1930s. Stored on the farm, it wasn't a cash crop so marketing wasn't a problem. In fact, families living on the farm used all of the crop the following summer. We called it a crop that relied on nature to ice our summers.

A Cool Harvest

Every year in January and February, the crew would harness up a team of Belgian horses and brace the cold, which was

JANUARY AND FEBRUARY ice harvesting took place on this calm and peaceful 18-acre lake during the farm's earlier days.

COLD WEATHER HARVESTING. Many wagon loads of ice were cut out of the lake and were transported by teams of horses to the icehouse each winter.

often far below zero. It wasn't fun work, but the crew knew this ice would save them come summer—when it was scorching hot.

For working in these icy conditions, the horses were shod with sharp shoes featuring pointed calks. This kept them from falling or slipping on the ice. Plus, these shoes really helped the horses's confidence and meant they weren't afraid to work on the ice.

After removing the snow from a small area of the lake with the horses and a scraper, the crew would hook a horse to the ice saw and cut a groove several inches deep across that portion of the ice they wished to cut. With four sharp knives on the ice saw, they would repeat this trip several more times to cut the ice to a depth of 7 to 8-inches.

The ice saw had a guide on it which would enable you to cut a straight groove every 30 inches.

After several rows of ice were cut in one direction, the crew would use the horse-drawn saw to cut the ice crossways. This would leave cakes of ice that measured 30-inches square and were as thick as the ice in the lake. Because these cakes weren't cut all the way through, the crew and horses could still work safely on the ice.

"The key was to keep the four knives sharp," says Dad. "The knives would usually cut 2 or 3-inches deep at a time. So if the ice was 10-inches thick, you wanted to run the saw over these grooves three or four times. That would give a cut depth of about 8-inches and then you could take a spud and later break off the last 2-inches of ice. In the meantime, the ice was still strong."

Colder In Old Days?

Dad says the lakes seemed to freeze deeper in the old days than today. "You would generally be cutting heavy ice cakes that were 30-inches square and 12 to 20-inches thick," he recalls.

After the cuts were made across the frozen ice, the next step was to take a long-bladed handsaw and slice long sheets of ice. After the crew had these long cakes cut, the crew placed a score mark on a whole row. Then all they had to do was take an ice spud, break off individual cakes along the scored lines and float them over to where the sleigh was parked.

The real trick, Dad says, was to keep from falling in the frigid cold water while you worked on the ice in January and February.

It called for fancy footwork on the part of the crew and the ability to jump from one island of ice to another without tumbling into the treacherous water.

The next step for the crew was to set up the ice chute. Dad says the crew tossed one end of the chute in the water and chained the other end of the chute to the sleigh. Using pike poles, the crew floated cakes of ice across the lake to the chute and pushed them up onto the sleigh.

While these ice cakes were certainly heavy, two men with pike poles could load plenty of ice without much difficulty. Once a cake of ice was on the sleigh, another crew member would grab each cake with a pair of ice tongs and set it on edge.

Okay, Let's Go!

Soon after, another sleighload of ice would be on its way to the farm's icehouse.

Our icehouse was actually the northern portion of the old woodhouse located behind the "big house." In later years when ice was no longer cut from the lake each winter, this area became a garage for my Grandfather's cars.

When the team arrived at the

THE WEATHER USED TO seem much colder than today and the lake would often freeze to where ice cakes were 20-inches thick.

ice house, the cakes were quickly unloaded and stacked flat on top of one another.

The crew was always careful to leave space around the outside of the cakes so none of the ice touched the double-boarded walls of the building.

If the ice touched the walls, it would have meant faster melting in warm weather. Any open space between the ice cakes was always filled with sawdust, which served as natural insulation to help keep the ice from thawing.

Cold Storage

Surprisingly, the ice could be held in this miniature blockhouse all through the summer months with remarkably little shrinkage. Double-boarded sidewalls and insulating the ice with sawdust were the real keys to keeping the "crop" in good condition. Everyone also was careful not to leave the door open very long when you went to get a cake of ice.

Dad says the crop of ice would keep in this insulated ice house from mid-winter through the following late August. Finally, small amounts of the remaining ice would slowly melt away.

When hot weather arrived, Grandma Norah would have Dad go to the ice house and dig out a cake of ice. After washing off the sawdust, the cake would be placed in the icebox to keep food cool during the summer months. Those were the days when foods were kept cool with ice—before electric refrigerators came to rural areas.

As a youngster in the late 1940s, I remember an ice box that still relied on real ice to keep food cool at the hired man's house. And when there was the occasional time when the electricity failed because of a midsummer storm, this old ice box would be in great demand once again by all of the farm families!

Dad says one of these old-fashioned iceboxes could gobble up two of these slabs of ice in 24 hours without even working up a sweat.

"We never used any ice until

Ice Cold Water

Working as a youngster on the farm's ice crews, Dad always had a fear of falling into the icy cold water. Every year, a few people in the area would fall into one of the half dozen lakes where ice was cut each winter.

"Once in a while, a horse would fall into the icy water," he says. "When that happened, one of the members of the ice cutting crew would jump into the cold water and hold the horse's mouth shut so the horse wouldn't drown. After a few seconds, the horse would float to the surface and the men would have to figure out a way to get the horse out of the water."

Dad remembers one early winter day when the farm crew decided to cut ice out of Dark Lake. The ice in the much bigger Dennis Lake where they normally cut much of their ice was not yet hard enough to work on.

"Unfortunately, the crew left their tools on the ice over night. When they came back the next morning, the ice had melted and the tools were sitting at the bottom of the lake," he says.

"Since that lake is more than 125-feet deep, there was no way of getting the tools back. More than 60 years later, there's still a full set of ice cutting equipment sitting at the bottom of that lake.

"It could have been much worse. Instead of the tools sinking to the bottom of the lake after the thin ice collapsed, we could have been in the water with the horses.

"That was an unusual year. The ice eventually really froze extremely thick. I remember we didn't get much ice harvested that year."

LATER USED AS A GARAGE, the old icehouse was located behind the big farmhouse shown at far right. Packed in sawdust to stay frozen, cakes of ice were stored for use until the remaining chunks finally melted away in late summer. It was critical to store the cakes of ice so they didn't touch any of the sides of the double-boarded building.

the weather really started to get warm in late May or early June," he adds.

"Sometimes on a hot summer day, my Mother or Dad would chop up an ice cake and make homemade ice cream. It was always a treat. If I helped with the cranking of the old-fashioned ice cream maker, I got to lick the dasher—rich with gobs of firm, succulent ice cream."

On a hot, humid July day, Dad and his friends always knew they could go to the ice house for a small piece of ice to lick or rub on their backs in an effort to keep cool. It was a great way to cool off after bringing in the latest load of hay or spending a hot afternoon shocking wheat.

Ice Cutting Mecca

Actually, the Lake Orion, Michigan, area where our family farm is located was a winter ice-cutting "hotbed" for many years. Dad remembers huge 40 or 50 foot tall ice houses located near Long Lake, Buckhorn Lake and Lake Orion which existed well into the late 30s.

"There used to be a railroad siding in downtown Lake Orion where crews would load ice during the summer months to ship to Detroit," he says. "When you went to town, you used to see teams of horses pulling wagons loaded with ice all summer long.

"The streets were always wet from dripping ice even when the sun had been shining for several days. This was really a big business for the railroad and the ice companies.

"The Franklin Settlement Camp land on Long Lake was originally purchased and used as an ice cutting operation. Later when the ice company quit cutting ice in the winter, they gave this land to a Detroit-based organization to start a summer camp for underprivileged big city youngsters."

Pets Add Lots To Farm Life

A FARM can have hundreds of acres of crops in the fields, pigs playfully rolling in the damp mud, cattle and sheep gathering under the peaceful shades of pasture oaks and a gold-colored rooster standing attentively delivering the morning welcome from the fence post.

But to me, no farm is complete without one of its unsung, yet integral ingredients—pets.

Farm pets have been a part of Lohill Farm for as long as anyone can remember. Although their jobs on the farm may seem minimal, pets played a big role in molding the personality of the farm, as well as bringing smiles and laughter to everyone—from a young child's first kitten to the eldest of generations who saw a new dog "get in the ring" with one of the farm animals for the first time.

Pet Family Tree

Dad says a collie named Prunes was the first farm dog he can remember. Prunes was adept at bringing in the cattle from the field all by himself, and was the only dog in Lohill Farm history that could herd cattle without any help.

Prunes was also the only dog ever permitted in the front parlor of the big house and he took full advantage of the house privileges, sleeping in comfort behind the old cookstove in the kitchen. He also was quite an entertainer.

"His acting ability when the

SADDLING UP. Grandpa Lessiter saddles one of the ponies for Susie to ride during her summer vacation at the farm.

The Cow Terror...

Muffet was Janet's dog and he loved to go with Dad to round up the cows at milking time. But he had one bad habit.

While scurrying to round up the cows, Muffet liked to grab hold of a cow's tail and swing back and forth as the cow ran for the barn.

Dad used to get really disgusted when the cow, with Muffet still hanging on to the tail, arrived at the barn completely worn out and all stressed out prior to milking time.

FARM DOGS OF 40S AND 50S. Scotty, shown at left pausing from talking on the telephone, was Frank's first dog. He later had to be put to sleep when Scotty bit the hired man's wife.

Below, Janet and Shirley Robertson play with Muffet and her very first puppy in front of the big barn.

DINGO WAS HIS NAME. A car accident left this dog with only three legs and one eye, but 3-year-old Andy begged his parents not to put his friend to sleep. Dingo turned into one of the best farm dogs the family ever had, always a true friend to Andy as they grew older together. Once, Andy was attacked by a turkey and Dingo fought the bird to keep Andy from being hurt.

THE PUPPY PALACE. Uncle Neil bred beagles for field trials and developed several American Kennel Club field champions. The dogs summered in cages in the pasture and spent the winter months in the old dairy barn.

GOING TO THE DOGS. At 14 years of age, Debbie is shown at left with one of Uncle Neil's beagle pups. Later, Debbie and husband Mark show off a later crop of farm puppies to two of their children, Molly and Alex.

old pump organ played was worthy of the best amateur hour performance," remembers Dad.

"Prunes would throw back his head and howl as long as the music continued. If you stopped playing, he'd quiet down, but as soon as you started, he'd join back in on the chorus again."

THE CAT FACTORY. Like most farms, there was a number of cats and kittens roaming around the barns. They kept mice numbers down and were always playful with visiting children. Several had no tails, the result of a dairy cow stepping on them.

Dad also remembers two large black and white cats his grandmother kept in the house. Aptly named William and Mary, Grandma pampered these cats as if they were the King and Queen of England.

"Grandpa thought they were real haughty and vain until William began cuddling with him on his lap," Dad says. "These were

NOAH'S ARK. Since Uncle Neil hadn't grown up on a farm, he quickly developed a keen interest in all kinds of farm animals. He soon was raising a variety of animals, including the baby goat which Katie and Andy are admiring at right.

Below, Mike really had his hands full with this very playful Lessiter farm dog.

TAKE IT SLOW. Susie takes a ride on one of the ponies.

the cats my grandparents enjoyed during their old age and they became privileged members of the household."

Later, a stray, shaggy little black and white mongrel (part terrier) wandered onto Lohill Farm one day and remained, capturing the hearts of the family. "He was surely no watchdog, but this dog we named Rags sure became a close friend," recalls Dad.

Rags was followed by Sport, a brown and white hound whose sad eyes and drooping ears belied his frolicking good nature.

SUMMER FUN. When the four grandkids came from Wisconsin, riding through the fields was always fun.

FREE PONY RIDES. It doesn't matter how old you are, Susie, Mike and Pam all enjoyed taking a ride on the farm's ponies.

On several occasions, he distinguished himself with disastrous run-ins with skunks.

As a young child, I had a collie named Scotty. After he bit the hired man's wife, we had to get rid of him.

In later days, we had a short-haired terrier named Muffet, a loyal and intelligent pup who grew especially attached to the females of the family. And after him came Jeff (we'll get to him later) and then Rex, a playful mutt who got himself into hot water by stealing eggs right out of the nests in the hen house.

They Worked, Too

These pets earned their keep. Cats had a lot of responsibility since without them, mice and rats could eat the feed or infect livestock with disease. But our cats always did their job and even brought back dead mice all the way to our driveway to show off their trophies.

In addition to helping with the cattle, the dogs served as "burglar alarms" protecting their home territory. I've seen foxes and raccoons retreating for shelter many times after one of the dogs caught a glimpse of them sneaking around the barns.

Although you can never really tell what effect they had on thieves, we were never robbed. Our dogs undoubtedly deserved some credit for that.

Loyal Dingo

As the area started to boom with newcomers from the big

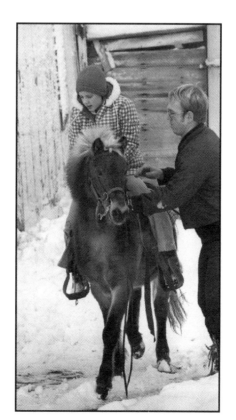

CHRSTMAS RIDES. Even though it was cold and snowy, the kids always rode the ponies when coming back home at Christmas. At left, Neil saddles a pony so Debbie can ride. One Christmas, Susie found herself bucked off in a snow bank with the pony falling on top of her. Everyone laughed, but she didn't think it was funny.

THE COWBOY. Andy, the little buckaroo, loved to ride his favorite pony or horse and listens intently as his Dad, Neil, gives him riding advice.

city, traffic increased greatly. Many of these animals, which long had all the freedom they wanted, died alongside the road. They were sadly missed by all of us.

Many years ago, a dog that belonged to my 3-year-old

DOWN MEMORY LANE. It was always fun to ride Chester down the old farm lane for nearly half a mile.

GREAT FUN. Susie and Kelly enjoy a ride on one of the horses. In earlier days, the only horses were Belgians used for field work. At right, Andy rides the Lohill Farm range.

nephew, Andy, was hit by a car, losing an eye and immobilizing one of his legs. After endless tears and heartfelt protesting, Andy convinced his parents and the vet not to put his dog, Dingo, to sleep. With only one eye and bounding across the field on three legs, Dingo wasn't going to win any beauty contests, but he was the happiest and most loyal dog I've ever seen.

Shortly after he healed, a viscious tom turkey cornered little Andy in the barn and was scratching and gnawing at the boy until Dingo saved the day, biting the turkey and keeping Andy from any further injury.

That dog hardly ever left Andy's side until it died of old age 11 years later.

Funny Times

Pets created some of the most hysterical moments I remember on the farm. It's amazing the messes which cats and dogs can get themselves into, ranging from trying to show another farm animal who's the boss to just being in the wrong place at the wrong time.

One of the more memorable sights came after Dad and I finished milking one day. After we were finished, we used to throw the gauze milk strainers to the cats and dogs gathering around us so they could lick off the excess milk. A couple of times, an over-indulging dog would swallow the entire gauze filter. You can imagine some of the laughs around the barnyard when we saw one of the dogs trying to "pass" the gauze the next day.

When I was older, we had an English springer spaniel named Jeff, the only purebred dog ever to roam the farm. One afternoon Dad was surprised to get a call from the county dog pound.

It seemed Jeff, who frequently disappeared on his own adventures, met up with another dog that morning and followed my young sister, Janet, to Webber School.

Apparently at recess, some first graders put down their sack lunches while they played, and Jeff and his "buddy" devoured the lunches, paper sacks and all. The two dogs ended up at the pound and it cost Dad $8 to get him out.

The next dog was a border collie named Mike. But he didn't stay long on the farm—soon getting hit by a county road commission grader.

Then And Now

Pets were usually seen everywhere across Lohill Farm. My nieces Trista and Katie had several calico cats and my brother-in-law has raised and trained beagles for over 20 years. It seems like there were always at least a few beagle pups romping on the front lawn in front of the big barn.

How About A Scratch?

Then there was the stout bulldog named Fanny. As you might guess from her name, she sat down in front of you and stuck her hind quarters up in the air to be scratched.

Since I was a kid, dogs, kittens, guinea pigs, ducks, rabbits, ponies, horses, birds, fish and just about every other pet imaginable have called Lohill Farm home. They certainly added a lot of fun and friendship to our lives over the years.

Having pets around keeps things interesting and our black Lab is no exception. Not too long ago he was outside playing when two ducks splashed into a huge puddle in our backyard. As you probably know, Labs are spectacular hunting dogs, so I carefully watched his reaction.

But what did the dog actually do? With his head down and tail between his legs, our panic-stricken dog scampered toward the door and wouldn't stop scratching until I let him in.

Man's best friend sure can put a smile on your face.

FARM PETS ARE GREAT. It seems like everyone always got plenty of enjoyment out of the various pets at Lohill farm, such as cousins Trista and Debbie.

62 Years Without Ever Missing A State Fair!

FROM THE TIME Dad was a child and up until the time he was in high school, his father and uncle were deeply involved in the purebred Shorthorn beef cattle business.

In fact, he took in his first

footsteps of their father (my Great-Grandfather Lessiter) in the Shorthorn business.

Some years before 1900, the family started showing cattle at the Michigan State Fair and a

number of other Michigan county fairs. They were highly successful showmen over the years, as evidenced by the hundreds of winning ribbons later made into a quilt by my great

> *"Showing Shorthorn cattle was long a Lessiter family tradition..."*

Michigan State Fair at 2 months of age!

Grandpa Frank and his brother, Floyd, followed in the

DECADES OF TRAVEL. Dad says this old show truck made the trip to hundreds of local and state fairs over the years.

grandmother and given to me when I was a youngster. There is also a silver loving cup, still in excellent condition, which was awarded to our family for showing the Grand Champion Fat Steer at the 1907 Michigan State Fair.

Incidentally, the Michigan State Fair is the oldest state fair in the country, having started way back in 1848. As a kid, I remember taking in the big 100th year celebration.

The original location for the state fair for its first 60 years was on the north side of downtown Pontiac which was much closer to the Lessiter family farmstead, being only 10 miles away.

The Michigan State Fair moved to its current location south of Detroit's Eight Mile Road in 1912. Over the years, there has occassionally been talk about the pros and cons of moving the fair to a more central site in the state, but nothing has yet come of the idea.

Many Exciting Fairs

Dad still has fond recollections of the early day fairs attended with his father and uncle.

"The first sign of the fair season was when my father and uncle (who farmed separately yet lived only half a mile away) selected those Shorthorn animals they would be showing," he says. "They would start feeding them in a darkened barn to obtain that perfect finish and shiny coat that often led to success in the show ring."

For several months prior to the traditional late summer fair season, the show cattle's daily diet included ground corn, oats and barley mixed with molasses and water. "I remember the flies that always flocked into the barn to pick up any spilled molasses," Dad says.

Shipped By Rail

Since our farm was located 30 miles north of the Michigan State Fairgrounds in Detroit, it was necessary to ship the show cattle by railroad.

"Shortly before fair time, my

A PRIZE SHORTHORN BULL. Photographed in 1918, this bull was among the farm's outstanding Shorthorn sires.

"Grandma always had her work cut out when show season rolled around..."

father would go to Oxford to order a large stock car and pay the freight to the Detroit fairgrounds," says Dad.

"After the railroad left the stock car on the Randall Beach siding which was 2 1/2 miles from the farm, my father and uncle would build an overhead platform in one end of the car to store feed, feed boxes, pails, forks, trunks and provide a place for the men to ride when traveling. There would be plenty of space under this 'upper deck' to tie calves, heifers, cows and bulls."

Dad says Grandma also had her work cut out for her at this time of year. She washed cattle blankets, fixed bedding and clothes for the men to take to the fair and packed the show trunks with necessary items, including plenty of time-tested home remedies just in case someone got sick at the fairs.

"She always packed a remedy which we called 'pain-killer'—a must to cure the dysentery we usually got from being unaccustomed to drinking Detroit's city water," Dad recalls.

When the day finally arrived for shipping the cattle, it meant

BEAUTIFUL TROPHIES. The vase on the left was for exhibiting the grand champion fat steer at the 1915 Michigan State Fair.

The center pitcher was presented for showing the 1909 grand champion fat steer and the pitcher on the right was for a champion Shorthorn female.

getting up extra early. Calves were loaded into a horse-drawn stock racked wagon for hauling to the railroad car. Older cattle were driven down the road to Dad's uncle's place where they would be joined by the Short-

---

"The 30-mile train trip normally took 8 hours..."

---

horn cattle he was shipping to the fair. They would then proceed to drive 20 or so Shorthorns to the railroad siding for loading.

Pop Bought Pop!

"I always remember the Randall Beach grocery store next to the railroad siding with fond memories," says Dad. "After the last cow and calf were loaded in the stock car, your Grandpa would always treat us to soft drinks at the store.

"I also remember how excited I was when I convinced your Grandfather I was big enough to go along in the stock car for the ride and help take care of the cattle at the fair.

"That first time, I quickly learned stock cars don't ride like passenger coaches on a train. I can still remember how scared I was the first time the stock car rocked severely as it rounded a sharp curve in the tracks."

Even though our farm was only 30 miles from the Detroit fairgrounds, the trip normally took 8 hours. This was due to considerable switching of freight and cattle cars, changing tracks and waiting for trains.

Once the stock car was spotted at the Detroit fairground docks, the cattle were unloaded and led to the nearby beef cattle barn.

"Just like today, work at the fair consisted of washing, feeding, watering and grooming cattle in preparation for the all-important show day," says Dad. "There was also the continual stall cleaning and spreading of fresh bedding that had to be done."

After all the work was done, the family would shake out bales of straw and spread blankets in the alleyways or in an empty stall for their own night's rest.

"I always enjoyed all of the good eats at the fair—probably more than my parents and other

MICHIGAN STATE FAIR. These four photos, including the Shorthorn judging competition above, represent various scenes from the 1912 Michigan State Fair. Originally held in Pontiac, the state fair moved that year to its current location on Eight Mile Road along the northern limits of the City of Detroit.

At right an excitied carload of youngsters takes part in one of the daily parades held during the 1912 fair.

DRESSED TO THE HILT. Farm families dressed up for visits to the Michigan State Fair, including this 1912 gathering in one of the fair's delightful groves.

At left, farm families check the latest seed plots displayed at the 1912 Michigan State Fair.

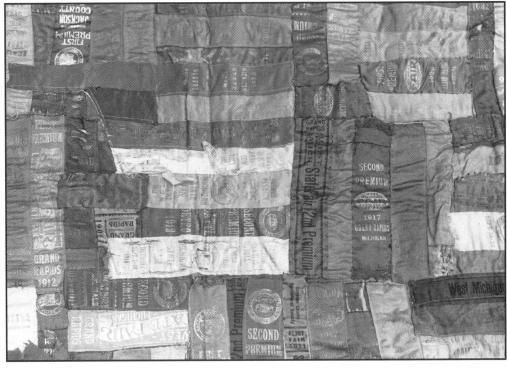

QUILT OF RIBBONS. For many years, hundreds of ribbons from livestock shows held during the first two decades of the 20th century sat in shoeboxes at the big farm house. Great-Grandma Wiser and Grandma Norah later sewed them together into the quilt you see here.

This special Lessiter family quilt was later given by Grandma Norah to her grandson, Frank. He later donated it to the Michigan State University Museum.

Below, Susie Lessiter examines the ribbon quilt which tells much of the history of the early-day Lessiter family livestock farming operation.

relatives," recalls Dad. "We always patronized one church-sponsored food stand which served meals year after year at the fair as part of their big yearly money-making project."

Other Fairs, Too

Besides the Michigan State Fair, the family showed Shorthorn cattle at the Jackson County Fair in Jackson and the big Western Michigan Fair in Grand Rapids. Those trips by rail took more

> *"State fairs were always in the blood of the Lessiter family members..."*

time and the family would often be gone for 2 weeks.

In 1924, when the beef cattle business was at a low ebb, my Grandfather and great uncle stopped showing Shorthorns. Since the farm was conveniently located in the Detroit milk marketing area which enjoyed good milk prices at the time, they switched to raising dairy cattle.

Fairs In His Blood

"But fairs were still in my

OLD MICHIGAN STATE FAIR. This photo of some of the outstanding Shorthorn cattle was taken at the Michigan State Fair in 1912. The fair was earlier located in what is now part of downtown Pontiac, Michigan.

Decked out in his fancy showring outfit, Grandpa Frank is shown at far right with one of the top finishing Shorthorn animals in that year's statewide event.

blood," recalls Dad. "When I was in high school, I went back to the Michigan State Fair and worked as a ticket-taker on the midway for a friend of the family. Each year, he contracted to provide workers to fair officials to double-check the honesty of the carnival workers.

"One year, they put me inside the merry-go-round to collect tickets. I would count the people riding the merry-go-round and make sure the ride operator had given me the correct number of tickets he had collected so the fair could get their proper cut of the income.

"With that ride constantly going around and around and the music blaring, I soon got sick.

"Another year I worked on the

THE HAY CUTTER. A sharp knife is missing, but long hay was sliced into short pieces with this unit. The 4-inch long hay was then mixed with grain for feeding the Shorthorn show string when housed at various Michigan fairs.

motorcycle thrill show where stunt drivers rode up and down the high circular walls at high speeds.

"After 10 days of collecting tickets on this ride for 9 hours a day, you certainly didn't want to see or hear another motorcycle for a long, long time."

When he was in college at Michigan Agricultural College in East Lansing, Dad worked at the Michigan State Fair as a livestock entry clerk. And he continued to do this for several years after graduating.

Enthusiasm for fairs still runs deeply in our family. In fact, both my sister and I worked as Michigan State Fair entry clerks during summer vacations from college. And we both showed dairy cattle at many 4-H fairs.

Many Years Of Fairs

"From the time your Grandmother took me to my first fair as a baby until 1969 when your Mother had major surgery during fair week which kept me at

"I had a perfect Michigan State Fair attendance record for 62 years..."

home, I had a perfect attendance record of 62 years at the annual Michigan State Fair," concludes Dad. "I've also attended a number of fairs since that time.

"I still feel the same sense of excitement I had as a young boy when I go to the fair. I hope the day will never come when county and state fairs are relegated to the past."

A Shorthorn Calf?

Grandma Norah always wanted to buy her grandson, Frank, a Shorthorn calf as a 4-H Club project.

She wanted him to follow in the footsteps of previous generations of Lessiter family members who raised and showed Shorthorns.

"In the mid-1950s, you needed a calf to show in 4-H and she said she would spend up to $700 on a Shorthorn calf for you," Dad recalls. "That was a lot of money for a calf."

"You wanted a Holstein calf instead. That was it—she didn't offer to buy you a Holstein calf. The Shorthorn calf was the only deal."

Blossom, The Sad And Lonely Calf

By Donalda Lessiter

THE BEAUTIFUL calf stood all alone against the pasture field wire fence on the sweltering late summer day.

Away from the rest of the calves, she wandered up and down the fence line, hoping to catch a glimpse of the little farm girl she loved so very much.

She couldn't eat, drink or enjoy the company of the other cattle. She just longed for Janet Lee.

A Beautiful Calf

This is the story of the sad and lonely calf named Blossom. She was such a beautiful little calf with her clean and shiny coat of black and white hair. She had big beautiful calf eyes and dainty little calf feet with polished hooves. But she was about the unhappiest little calf you've ever seen.

She had not always been a sad and lonely calf. You see, she belonged to a little girl named Janet Lee who lived on a big farm with her mother, father and big brother. And Janet Lee loved this little calf very much.

From the first time her dad showed it to her minutes after its birth, Janet Lee loved it and wanted it for her own. In fact, she wanted the little calf so much that she coaxed her daddy, Farmer John, into letting her have it.

Well, Farmer John, who loved Janet Lee just like all daddies love their little girls and boys, couldn't refuse her. Besides, if

---

A MESSAGE FOR CHILDREN...

This children's story was based on the "real life" troubles of my sister's 4-H dairy calf and sends a message to children about the jealousies of youth.

One of Mom's greatest pastimes was writing stories such as this one for each of her seven grandchildren. And her subjects usually came from quirky experiences that happened at Lohill Farm. This story is based on the experience of one of my sister Janet's 4-H dairy calves.

My mother, a former schoolteacher, used this story not only to entertain her grandchildren, but also to teach them a valuable lesson about life and dealing with others, both on and off the farm.

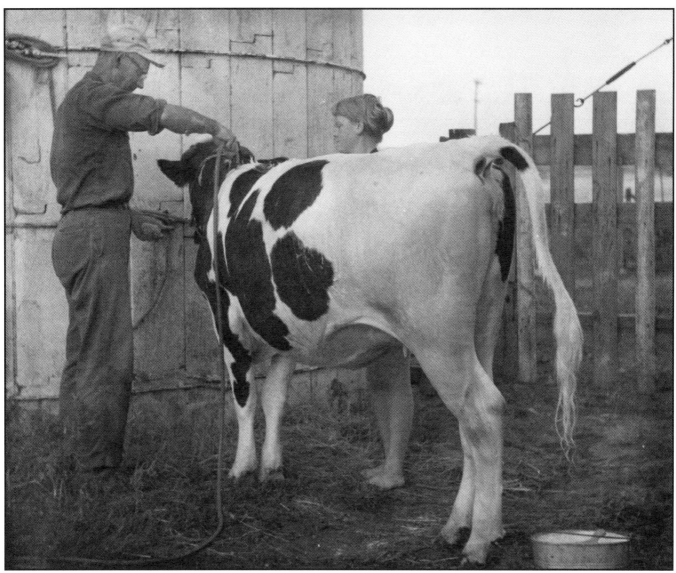

GETTING READY WITH A BATH, John Lessiter helps his daughter, Janet, give her yearling Holstein, Blossom, a bath in preparation for the upcoming Oakland County 4-H Fair. At far right, Janet gives Blossom her last bath at the wash rack prior to show day at the fairgrounds.

Janet Lee was going to show a calf at the 4-H Club County Fair, he'd have to give her one. So that is how the little calf became Janet Lee's.

And guess what name Janet Lee had picked out for the little

> *"The calf began to 'blossom' like the lilacs by the side of the house..."*

calf? "Blossom!"

Janet Lee said after she brushed and petted the calf, she began to blossom just like her mother's spring lilacs by the side of the house.

Lots Of Love

Never before had a little calf been so well cared for. In the big barn she had a stall of her very own where Janet Lee put fresh straw every day to make a nice clean bed for Blossom to rest in when she was tired.

Janet Lee gave the little calf feed and water each day so Blossom would be healthy and pretty. If a little dirt did get on her coat of shiny smooth black and white hair, Janet Lee scrubbed her just like mothers scrub their children when they get dirty.

Each day, Janet Lee took Blossom out of the barn and walked her. At first, Blossom was afraid. She would jump and run and try to get away from Janet Lee. But she would just hang tightly on to the rope halter and pretty soon Blossom began

to like the daily walks. And she learned to do just as Janet Lee told her to do.

Now and then, during her walks through the barnyard, Blossom could see the hilly pasture beyond the wire fence. There, she could see all the other calves eating and drinking together. She wondered what it would be like to have friends of her own kind.

Well, needless to say, when county 4-H fair time came early in August, Blossom was indeed a beautiful show calf. Janet Lee was so proud of her. This was especially true when they went out in the show ring where all the other boys and girls were showing off their calves, too.

Blossom behaved so well. When the judge paused beside Blossom, she held her head up high without being reminded by Janet Lee, and she took those slow dainty steps Janet Lee had so much trouble teaching her.

A Proud Duo

When the judge gave Janet Lee a blue ribbon and hung a flowered ribbon around Blossom's neck, Blossom licked the judge's face. It was hard to tell whether she or Janet Lee was the proudest. Everyone said such nice things about the little calf. And all the little girls and boys who came to look at the animals at the fair wanted to stop and pet her.

Oh no, indeed, Blossom was not always a sad and lonely calf. It was when she came home from the fair that it all began.

Farmer John told Janet Lee: "Now that Blossom is older and you are through showing her at the 4-H fair, I think it would be a good idea to put her out to pasture with the other young calves and let her learn to eat that good green grass. Plus, she should get more exercise until winter comes."

Janet Lee couldn't stop the tears from rolling down her face. She had spent all year with Blossom and had grown very attached to her.

Ever since Blossom was born, Janet Lee said hello to her every morning, and always made sure she had enough to eat and drink. After she got home from school, she would brush her and take her for daily walks around the barnyard. But she knew her Dad was right.

Life Can Be Tough

So the next morning, Janet Lee handed Farmer John the halter rope, threw her arms around Blossom's neck and said her goodbye.

Farmer John turned Blossom out to the field of grass right behind Farmer John's house where the family could keep an

"When Blossom ran over to the other calves, they didn't even bother to look up at her..."

eye on her and the other nine calves that had been there all summer long.

Blossom liked the taste of the fresh grass and she could hardly wait to tell the other calves, whom she hadn't seen since she was just a tiny calf herself, all about the 4-H fair.

But when she ran over to where they were eating, they did not even look up at her. And when she tried to talk to them, they walked away.

When she tried to follow them, one calf lowered his head and bumped her with it so hard that it hurt. Another one mooed: "That ought to teach you a lesson. Maybe now you will know when you are not wanted, Miss Blue Ribbon."

Poor Blossom!

She could not understand why the other calves did not want her around. She tried again to make friends with two of them and again they pushed her out of the way and ran her against the fence so hard that it smarted. One of them mooed: "That will teach you to strut and show off. We will fix your fine looks for you."

When Blossom asked what she could possibly have done to make them dislike her so much, the other calves would not even answer her.

So naturally, Blossom began to look sad and lonely. She missed Janet Lee and the things they used to do together. Before, she would see Janet Lee twice a day. But now it seemed she was all on her own.

When she got thirsty, she noticed two of the calves putting their heads down in a big tank over by the fence. Of course, she was used to her own drinking pail, but if this was the way the rest of the calves drank their water, she guessed she could too.

She edged up to the big tank to drink, but the other calves would not let her near it until they were all through. They left only a tiny puddle in the bottom of the tank and ran off without even so much as a friendly "moo." She sadly swallowed the last drops of the cool water.

Blossom noticed the other

DURING A MID-WINTER workout, Janet gives Blossom a good brushing and some well deserved special attention.

calves had settled in the shade of a big tree and walked over to join them. She mooed a friendly "hello," but they just ignored her and continued to chew their cuds.

When she laid down in the cool shade, they all got up and

> "Blossom had never been so sad and lonely in her entire life..."

walked away. One of them muttered, "Show calf indeed. Where are your friends now?"

Hurt, Hurt Feelings

You see, they were jealous of Blossom and her fine looks. Blossom had never been so sad and lonely in her entire life.

That night, Janet Lee watched Blossom standing on the hillside under the moonlit sky. Blossom lifted her head, saw Janet Lee in her bedroom window and tried to fake a smile.

The next morning, when Janet Lee came to the fence to pet Blossom, she saw how the other calves were treating her and how sad and lonely Blossom looked. She noticed how red and unhappy the little calf's beautiful soft eyes had become.

That night at supper, Janet Lee told her mom and dad how sorry she felt for Blossom and how the other calves were treating her. She asked if she could bring Blossom back to the barn.

Farmer John told her not to worry. "It just takes a little time for them to get used to her," he said. "It is just like having a new little boy or girl come to live in our neighborhood."

But the days went by and the other little calves treated Blossom no better. She just became sadder and lonelier. She was not brushed daily and the other calves made fun of her looks.

Finally, she began to stay away from them. Her head dropped and she no longer walked proudly. She was also getting quite thin and she was so sad and lonely.

The New Arrival

And then one day, Farmer John was away all day at a cattle sale. He came home with a big surprise. He had bought a fine,

---

"At first they were nice to her, hoping she would put in a good word for them with big and handsome Hank..."

---

big and handsome calf named Hank.

When this calf was turned out to pasture, the other calves all tried to make friends with him. That is "all except Blossom" who knew it wouldn't do any good anyway. No one wanted to be friends with her, did they? The other calves had made that quite clear to her.

So she just stood back and watched the others as they told big and handsome Hank where the greenest grass was and showed him the water tank and the cool spots in the shade.

But this time it was big and handsome Hank who would have nothing to do with the other calves, except to order them around. And if he wanted the choicest place to eat grass, he just took it. If he wanted a drink of water he took that too without even waiting his turn.

He just dared any of them to stop him from doing exactly as he pleased in the pasture. And the other calves were just plain frightened of him. So they continued to follow him around and did just as he ordered them to do.

All except Blossom. You see, it just so happened that big and handsome Hank liked her. He saved some of the best hay for her and kept trying to stand next to her.

Well, at least she had one "friend," and she began to feel less sad and lonely and started to look a little happier. But she longed for the other calves to like her.

And then it happened. One by one, the other calves began to gather around Blossom. At first they were nice to her, hoping that she would put in a good word for them with big and handsome Hank.

Before long, they found out they genuinely liked her. She was always so kind and gentle and very concerned about the other calves. She asked them about living on the farm and what summer in the pasture was like.

And when big and handsome Hank was unkind to the other calves, she defended them and talked back to him. She even asked him to be nicer. And she was so sweet that even he could not refuse her.

The Christmas Morning Question?

Should the farm kids be allowed to rush downstairs and open their presents from Santa or have to wait until after Dad was back in the house from doing the morning livestock chores?

He soon realized he hadn't been treating the other calves very nicely and said he was sorry to every one.

Well, the other calves were so ashamed of themselves for the way they had treated Blossom that they could not do enough for her now.

And Blossom?

By the time Farmer John moved the calves back into the barn for the winter, she was just about the happiest and least lonely calf of them all. She just mooed with happiness. And best of all, everyone was such good

---

"Now her happy self again, she was no longer the sad and lonely calf..."

---

friends. Janet Lee now fed them all every day and spent every afternoon in the barn with all the calves.

A True Champion

The other calves were now always interested in hearing about her adventures at the 4-H fair and hoped she would do better than ever next year. After all, they now liked her best of all and they were proud to have her in their stalls. And they knew she was a true champion.

So Blossom was no longer the sad and lonely calf. She was her old happy self again.

She held her head high and her eyes shone when Janet Lee petted her and whispered in her ear, "Just wait until next summer's fair time. We'll show them all what a real, true blue champion you are."

Going With Dad To The Detroit Stockyards

WHETHER WE HAD A cull dairy cow, an old sow or a few lambs ready for market, it always meant a trip for Dad in the pickup truck to the Detroit Stockyards.

Trucked Our Own

Other farmers in the area often sent animals to the yards with area livestock truckers or local livestock jockeys. But it was a long-time family tradition that we always accompanied our animals to market, whether it was by rail car in the earlier days of the 20th century or later by truck.

As a kid growing up at Lohill Farm, it was a fascinating half-day trip to go with Dad to the stockyards which were owned and operated by the New York Central Railroad system for many years.

In earlier days, railroads owned many of the country's

THE OLD YARDS. The old gates and fences are long gone from the Old New York Central railroad's location on Detroit's western side. But these were the pens used to bring in animals by rail and facilities used to ship cattle, hogs and sheep to eastern packing plants in the days when the railroad was still the king of livestock transportation.

stockyards since so many animals came and went by rail.

Yet by the time I started making the trips with Dad in the late 1940s, it was already plenty quiet at the Detroit yards. We did all of our livestock selling through the Michigan Livestock Exchange, a producer-owned cooperative still going strong today, although the methods of selling livestock for members have changed drastically over the years.

Even in those days, the Exchange was one of only two remaining commission firms operating in the yards—a vast change from the glory years when dozens of commission firms competed at the Detroit yards for the trading business of Michigan and northern Ohio farmers.

Rush Hour Traffic

After finishing the milking and eating breakfast, we'd get the animals destined for market loaded in the truck and start the roughly 40-mile journey south to the Detroit Stockyards.

Starting at the tail end of the morning rush hour traffic, we'd head down Baldwin Road to downtown Pontiac, drive along the beautiful three and four lane Woodward Avenue to the outskirts of Detroit.

Woodward Avenue, a divided

MILES AND MILES OF FENCES. In its heyday, the Detroit yards played a key role in the sale of livestock throughout the eastern area of the country. Before the Chicago Stockyards became the country's top livestock market, New York Central officials felt Detroit might take on that all-important role.

highway with a flower, grass and tree-lined boulevard, was the main thoroughfare running the 30 miles from Pontiac to Detroit. It didn't matter what time you made the drive, there was always plenty of traffic on Woodward!

Next, we'd head a few miles west on 8-Mile Road and swing south on Livernois Avenue, which would eventually bring us to the Stockyards area.

There was plenty of traffic to fight on this trip to the yards, especially once we got into the Detroit area. While it was only a 40-mile trip, it would easily be 2 hours before we made it to the yards.

The worst part was the tedious 10-mile trip down Livernois since there was a traffic light on the corner of every Detroit city block. There were hundreds of cars and trucks to avoid and thousands of busy Detroiters crossing the streets.

Bulls, Trucks, People

Those trips when we had a big bull in the stock rack-equipped pickup could get mighty exciting as they never seemed to adjust to the rhythm of the truck when it stopped for traffic lights. And it often meant a rough ride for everyone.

At the yards, Dad would back the pickup truck up to one of the

"By mid-morning, most of the excitement in the yards was already over..."

unloading docks—there must have been 40 of them. Since it would now be mid-morning, most of the excitement at the

MARKETS FOR EVERYTHING. There was always a competitive daily market at the Detroit yards for every species of livestock. Few animals ever spent more than a day or so in the busy yards before moving to market or to a farm for further feeding.

yards was already over. Most of the incoming livestock traffic at the yards took place in the wee hours of the morning—often before the sun was up.

After unloading and parking the truck, Dad and I would follow along as the animals were moved through the maze of battered old gates and alleyways to the scales.

After being weighed, the livestock would be moved through another maze of gates and alleys to the pens used by the Michigan Livestock Exchange commission men.

Sale of the livestock usually

WHEN TRAINS WERE KING. Over the years, hundreds of thousands of cattle, sheep and hogs moved through these railroad facilities at the Detroit yards to eastern packing plants.

didn't take long as a number of livestock buyers were mingling around and would soon make bids on our animals. Once the bid was accepted, Dad would know whether he got a fair price for the animals or not.

While he often felt he should have received a cent or two more per pound for a cull cow or bull, it wasn't like we were moving 100 head of fed cattle to market as some folks did.

These feeders could lose big, big bucks if the market turned a cent or two lower that morning and they'd picked the wrong day to bring in their cattle.

Thousands Of Gates

After the sale was over, we'd make our way back to the truck. Walking through the alleyways was no easy task as you had to open and close plenty of old wooden, sagging gates along the way.

To a young farm boy, it seemed like each one of them was sagging and had to be dragged across the cracking concrete.

Next, we'd pile into the truck and drive over to the southeast corner of the yards where the Detroit Stockyards Livestock Exchange Building was located.

It was headquarters for the commission firms, yards management, government market reporting services, livestock insurance agencies, restaurant and other essential stockyard support groups.

I can't remember for sure how tall this building was, but I'd guess it was four floors. We'd walk up to the second floor where the Michigan Livestock Exchange firm had offices and wait while they wrote out a check for the livestock. It was very unusual for Dad to head home without the livestock having been sold and the money stashed away in his pocket.

The Poultry Place?

Sometimes, we would take along a crate or two of old hens. On our way out of the yards, we'd stop across the road at a Detroit poultry firm and they'd buy the old hens.

Even in those days, the New York Central Railroad used very little of the acreage they owned around the yards.

The speculation was that the railroad had bought up land in earlier days with the intention of making the Detroit Stockyards into a market that would have competed with the huge Chicago Stockyards. But this wasn't to be, especially when livestock started moving by truck instead of rail.

You could see the old train

DOUBLE-DECK CARS. The ramp and platform at left was used to load upper decks of hogs and sheep on railroad cars for the trip east.

roundhouse for steam engines which was no longer being used.

More than once we went into the old train maintenance building where an unpainted furniture store rented space. One time, we bought a bookshelf for my room and trucked it home that day in the straw-bedded pickup truck.

Bag Of Burgers

By the time Dad and I had journeyed back north of 8-Mile Road and were heading north on Woodward Avenue toward the farm, it would be time for a late lunch.

It would be a special treat when Dad pulled into one of the old White Castle restaurants in Royal Oak for lunch. This was a fast-food restaurant before anyone ever knew of food places by that term.

Dad and I would sit at the counter and order up a "bunch" of their juicy hamburgers. I say "bunch" because each one contained only 1/2-ounce of beef and sold for 9 cents each.

We'd wolf down ten of the burgers between us and being good dairy farmers, polish off a couple of glasses of chocolate milk.

A Neat Time

These 5 to 6 hour trips were always a neat journey for me. They gave me a chance to see how livestock was sold in the big city stockyards and let me spend some special time with Dad. Of course, spending time with Dad was never a problem for farm kids in in those days since farm families lived and worked together all the time anyway.

It's been more than 45 years since I've been to the Detroit yards. The old pens and alleys are no longer there. The yards

219

COMMISSION HOUSE ROW. A number of livestock commission firms always competed fiercely with each other at the Detroit yards for the right to market the farmer's livestock.

OKAY, LET'S WEIGH THEM! Since everything was sold on the basis of price per pound, certified scales were a critical part of the stockyards.

were closed sometime in the 1960s as the traditional means of marketing in this country underwent drastic changes.

But these were jaunts to the old Detroit Stockyards in the pickup truck with Dad and the cows, hogs, lambs or chickens which I still fondly remember 50 years later. And I still remember those delicious White Castle burgers, too.

Bull Rings, Gilt Rings, Threshing Rings And Other Rings

MENTION THE WORD "rings" on a farm years ago and the thoughts didn't turn to diamonds, rubies or sapphires.

Instead, "rings" usually brought talk of bulls, boars, gilts, threshing machines and potato diggers. On other farms, it included other farm animals or a steel-wheeled or rubber-tired piece of machinery that could have a major impact on income.

At Lohill Farm, I remember several "rings"—some which were fairly organized and others which simply relied on neighbors helping neighbors.

Dad and a half-dozen other area Holstein breeders jointly owned several registered bulls when I was growing up. Each farmer would keep a bull

THE GILT TRIP. East Orion 4-H Club members raised bred gilts sponsored by the Pontiac Kiwanis to start hog projects.

through a breeding season and the bulls would be moved to different farms the following year.

This was before artificial insemination came on the scene, so all breeding had to be done the old-fashioned way.

The idea of the "bull ring" was to find genetically superior bulls and rotate them among the breeders. But it wasn't always easy to keep everyone happy, especially when one member made an unfavorable comment about someone else's favorite home-reared bull.

I remember a time when nobody was home and one of the bull-ring members delivered an ornery young bull to us. He let it loose in the barnyard instead of securing it in the bull pen.

It was a miracle the bull didn't break loose and head for the fields. And it certainly was no picnic trying to get the scared bull into his new surroundings without anyone getting hurt.

With fewer Holstein herds in the county, the local bull ring shut down in the late 1940s.

Some of the biggest boosters of dairy bull rings were college dairy professors. They saw these rings as a way to bring better genetics to dairy herds. Some extension workers wrote model constitutions and bylaws which could be adopted and used anywhere. I don't remember Dad and the other Holstein breeders having an official set of rules, but I do recall seeing notes from meetings in my younger days.

The Gilt Ring

Actually, the gilt ring was one in which I participated rather than Dad. It was a group of youngsters with a bred gilt traded among several 4-H Club members.

The original bred gilt was purchased by the Pontiac Kiwanis Club, and the idea was that a gilt would provide each 4-H Club member with a quick start on a pork production project.

Each youngster was to farrow a litter of pigs, then pass a gilt along to the next 4-H Clubber.

The project worked well and I was able to get several litters of pigs out of my gilts. Besides being a good 4-H project, the pig project became a good source of income and college funds.

The Threshing Ring

The folks next door at Lakefield Farms had a threshing machine and we always worked together to get everyone's small grain in the barn as quickly as possible. We'd spend a few days helping them haul grain from the field and getting it threshed.

Then they'd move the threshing rig to our farm and we'd repeat the process. With all of the available help, it usually only took a few days to wrap up our 50 or so acres of wheat, oats, barley or rye.

The Potato Digger

Dad also partnered with Frank and Neal Dowling on a potato digger. Each farm raised a few acres of potatoes and we'd trade the digger back and forth in the fall to harvest spuds.

The potato digger agreement finally fell apart because Frank's wife felt they were being taken advantage of (which definitely wasn't true). Not only did we store the digger in our machine shed, we kept it repaired and in tip-top shape.

Disgusted, Dad finally told them to keep the potato digger at their place and they could take care of it. Pretty soon, the days of sharing the potato digger and raising spuds came to a close.

Nothing Official

Unlike some rings, these were pretty much neighborhood groups without any heavy layer of official rules and regulations.

Other threshing rings started out as an informal grouping of neighboring farmers intent on helping one another. Some rings changed in time as they became more formal and even adopted constitutions, bylaws and kept careful records to ensure that everyone was treated fairly.

While farmers started the original ring movement in many areas, extension workers later stepped in to impose order and precision.

Sometimes this highly regulated "ring" concept worked, but sometimes it didn't.

MACHINERY RINGS. Since each farmer only raised a few acres of potatoes, several neighbors owned this potato digger together to trim machinery costs.

When Sundays Were Very Special

"AFTER 6 LONG hard days, it was always good to see Sunday roll around," Dad recalls from earlier days spent on the farm.

While city folks thought he must dread having to do livestock chores on Sunday, Dad was always glad those were the only chores he had to do on Sundays.

Dad's now in his late-80s, but he still really enjoys talking about the many good old days on the family farm.

"Sunday was always special in a number of ways," he remembers. "While your Grandfather probably didn't give Sunday much thought until late Saturday afternoon, your Grandmother always started getting ready for all

SUNDAY FAMILY TRADITION. Posing for the camera while waiting for Sunday dinner is Grandma Norah with her only granddaughter, Janet.

SUNDAY FARM PICNICS. Grandma Norah and Grandpa Frank (second from left in the bottom row) frequently held Sunday gatherings such as this one for their friends. Despite the warm summer weather, wearing your Sunday best clothes was always essential. Both of my Grandparents really loved to entertain in their big farm house.

the Sunday activities no later than Friday."

Dad says that's when she cleaned the big house thoroughly and left only the kitchen floor to be scrubbed on Saturday morning. Once that was done on Saturday, she spent the rest of the day baking and cooking.

"Oh, it was really something to come in from outside and take in all those delicious fresh-baked

RESTFUL SUNDAYS. While the adults always visited after the Sunday meal, the younger folks often took long walks back through the fields to enjoy the farm's own private 18-acre lake, the wildflowers, the wildlife and the growing crops.

Favorite Early-Day Farm Foods

WHEN DAD was growing up, there were a number of special treats enjoyed by all, including a few favorites of the Lessiter family to this day. Here's what he remembers most:

Boiled Apple Pudding: This was always my Grandfather's favorite dessert.

To make it, my Grandmother wrapped dough around peeled apples, sliced them and then cooked the apple slices for 2 or 3 hours in a steam bath. After they were cooked, brown sugar and homemade cream were put on top. This pudding was a real delicacy—only served every 6 months or so...and only if fresh apples were available.

Creamed Codfish: A regular weeknight supper every 2 weeks or so, Dad says it was always delicious.

Grandma would buy codfish in big wooden boxes and store them for months in the cellar. She would make a cream sauce from the codfish, then serve it over boiled potatoes. I remember eating this myself on many occasions as a kid at her house.

"The creamed codfish was a real treat I enjoyed," says Dad. "But after I married your mother, I found she hated it. But your Grandmother would still serve it when your Mom was at a meeting and I ate with her.

"My cousin, Bruce, used to think creamed codfish was a meal fit for a king. He even liked it better than eating fried chicken—I didn't go quite that far!"

Hermits: Made with hickory nuts and fruits, Dad fondly remembers the fact that these cookies were always small.

"Your Grandmother almost always baked cookies in only one size—extra large," he says.

Sugar Cookies: Grandma's sugar cookies were always at least 3 1/2 inches in diameter. "They always featured one big seeded raisin right in the middle—none of those little seedless raisins like you get in the supermarket," he says.

Angel Food Cake: Even though Grandma never really liked these cakes herself, she was known throughout the area for her delicious angel food cake. She'd take a dozen egg whites and whip up an angel food cake several times a week. The left-over yolks always found their way into a sunshine cake—the kind of cake Dad, Grandpa and Grandma liked best. But not me—I always liked her angel food cakes best. In fact, she would often bake one just for me.

Even when she was well into her 70s, I can remember her baking one or more of these cakes if people would be dropping in or if a party was

GRANDPA LOVES CARROT CAKE. Another Lessiter family favorite food, shared here with granddaughter Kelly, is carrot cake. In fact, John eventually became a top-notch cake baker.

being held. She would sometimes frost these cakes while she wouldn't at other times. Either way, they were always delicious.

If Grandma was going to a party or would be bringing something to a bake sale, everyone in the area knew it would always be an angel food cake. They just took that for granted when making their plans.

After Supper Treats: As is still the case with the Lessiter family, one of the great after supper delights was a couple of poppers of popcorn. And we always add a little butter and salt to make it tasty—regardless of what today's health experts might say.

"Back when I was a kid," concludes Dad, "we used to have relatives who liked to put milk on their popcorn. But I couldn't stand it...in fact, I absolutely hated it."

LESSITER DESCENDANTS. Grandpa Frank and Uncle Floyd were brothers and were the first two Lessiters born on the home farm. Each later operated their own farm, located 1 1/2-miles apart. Dressed up for a Sunday party, Dad is shown at right with Bruce, the only son of Uncle Floyd. In the center is Mary Jane Reese, Bruce's only niece.

aromas," Dad recalls. "Loaves of bread, pies, cakes and cookies were soon being stored in the pantry off the kitchen, all waiting for Sunday's dinner and supper.

"There were jams and spice bins in the cupboards that added to the mixture of scents, too. And if that wasn't enough, there was always the basement storeroom where milk, cream, eggs, canned goods, crocks of pickles, home-grown vegetables and home-made butter were kept.

"Many were the times when I was sent to that cellar storeroom to fetch something for a meal and I lingered a lot longer than I needed to, just to enjoy all those great aromas floating around."

Church Time!

After the livestock were fed and watered on Sunday morn-

"As a kid, I always looked forward to the friendly pat on the head by the old minister..."

ing, the family would start preparing for church and Sunday school. This meant dressing up for everyone, including hat, gloves and the very best dresses for my Grandmother, Great

WEEKLY TRADITION. Shown here at Mom and Dad's 40th wedding anniversary party, Mom and Mary Parker almost always had afternoon tea together on Saturday afternoons. Mary was 100 percent British and she loved to talk about the week's happenings over afternoon tea.

LITTLE SAILOR BOY. Mom and Frank were all dressed up for church and Sunday dinner in this early spring photo taken by Dad. Note the horse-drawn one-bottom moldboard plow sitting behind them. With Uncle Sherm serving in the U.S. Navy on a destroyer in the Pacific Ocean during World War II, Frank was outfitted in this Navy uniform—just like the one which his favorite uncle wore.

Grandmother and any lady visitors. (Even though Dad was an only child, the big house had plenty of room for visitors—with six big bedrooms).

Except in winter, the trip to the little country church was made by horse and buggy. During the snowy winter months, the trip was made with the horse pulling a sleigh.

Going to the church in the winter was always much warmer than the long trip home. Heated bricks were placed under everyone's feet before the sleigh left the farm.

While the family was in church, the horses spent the worship hour munching hay in sheds out behind the church. It was always a source of great pride to the men to have their horses well-groomed and buggies cleaned and polished.

With no nurseries, even the tiniest of babies were regular church attendees. For the restless child, pink peppermint candies carried in a pocketbook by one of the older members of the family would cure practically any problem.

"As a kid, I always looked forward to the friendly pat on the head by the dear old minister as he shook hands and greeted all his parishioners with real love as we left the church," recalls Dad.

Next, Sunday Dinner!

As soon as the family returned home, my grandmother would bustle around preparing the scrumptious Sunday dinner. This biggest meal of the week featured fried chicken plus all kinds of other good things to eat.

Mainly prepared on Saturday, there was always ample food available in case any number of visitors just happened to drop in for dinner.

Grandma would always have an angel food cake for dessert if she figured visitors would be at the dinner table.

Sunday dinner was recognized throughout the Lake Orion farm community as a day of visiting. The best china was laid out on a snow white linen tablecloth which replaced the checkered ones used the rest of the week.

Fun And Games

Sunday afternoons were free as no unnecessary work was done in the house, the fields or the barn. While the adults— including visitors—visited or

REPRESENTING FOUR LESSITER FAMILY GENERATIONS. Following one of the highly popular, mouth-watering and traditional Lessiter family Sunday dinners, Dad holds the highly-active 2-year-old Ryan, one of his great-grandchildren.

have prepared the traditional Sunday meal in the Lessiter household.

"All the time I was growing up, our Sunday night meal was always a mush supper," recalls Dad. "Mush made up the whole meal, but that was no problem

---

"I haven't eaten any mush in over 45 years and I really can't say that I've missed it..."

---

dozed in their chairs, the younger people often took long walks back through the fields to our own private lake.

After returning to the house for cookies and lemonade, the hand-cranked phonograph would be turned on and everyone would listen to the beautiful music.

Another regular source of Sunday afternoon entertainment was the family picture album. Dad says it was always great fun to turn the many pages and see all the photos time and time again.

By contrast, Dad's feelings were entirely different about the framed pictures of ancestors that hung on the parlor walls. Solemn and dignified, they constantly stared down on everyone.

"I never felt at ease with them and could never imagine how they ever could have been a part of me—although most of them I knew personally," he says.

Mush For Supper

After taking care of the evening livestock chores, Dad and Grandad would return to the house. My Grandmother would

since we'd always had a big Sunday dinner. We were never that hungry on Sunday night anyway."

Many of you probably don't know anything about mush and, in fact, I didn't myself. Dad says it was simply corn meal boiled in water. Then you added brown sugar, butter and sugar or milk after you sat down at the table.

"It was nothing fancy, but we ate it for supper on Sunday after Sunday for years and years," says Dad. "I enjoyed it in those days, although to tell the truth, I haven't had any mush in at least 45 years and I can't say that I've really missed it."

Putting On Your Sunday Best

JAB!! RIGHT IN THE RIBS! Wincing from the attack, I knew what hit me before opening my pasty eyes.

As "Amazing Grace" rang out from the organ, I turned to my wife sitting next to me and got that disgusted look from her for falling asleep in church again.

After the song was finished, the reader read an announcement about the Easter decorating committee, which was planning new ways to "dress up" the church for the special day. As we left the church, many memories about my old church came to mind.

"Our" Church

Our church played a huge role in Lake Orion's development and my parents were both long-time members. Mom went to that church all her life and my parents were married there. Dad recalls there wasn't anything there but the church building then, and after the ceremony, everyone gathered in the basement for cake and ice cream.

The church was built in 1872 on Flint Street, near the railroad that ran from Detroit to Bay City. "It was so loud when a train was going by that nobody could ever

125-YEAR-OLD CHURCH. This is the second site where the Lake Orion Methodist church has stood. It was moved to its present location in 1901.

ACTIVE MEMBERS. After Mom and Dad were married in the church on July 16, 1938, they became very active members of the congregation.

hear the preacher preaching," says Dad.

So in 1901, the parish members cut trees and put the church on top of the logs. With the help of some draft horses, they rolled it about four blocks down the street, lifted it up and put it on top of basement rooms built underneath.

Dad recalls a time as a youngster when church was sometimes a "spur of the moment" type of thing around the house. "Dad would just come in after tending to the farm animals some Sundays and say, 'Let's go to church this morning,' and we'd be on our way'."

Social Aspects

Mom was very involved—even at an early age. When she was in high school, she belonged to the Epworth League, a popular Sunday night program that was mostly a social group. They'd get together and talk, raise money for different causes, go to other churches for programs and do all kinds of things.

After they were married, Mom and Dad really became active. They were on several church improvement committees and were the sponsors for my sister Janet's youth group. We always had a group of kids at our place on Sunday nights.

Grandma Tarp was very instrumental in helping build the original fellowship hall for holding dinners, coffee socials and other community events.

Well remembered was a big party which Mom threw for Reverend Walker's retirement. "He wanted to move back to his hometown of Hadley, so I took the truck and moved him up there," Dad says. "A year later, he missed Lake Orion, so I drove back up and moved him back down again." As a kid, I remember making both trips with Dad.

Mom worked very hard on getting a historical marker for the church, and in 1973, the state

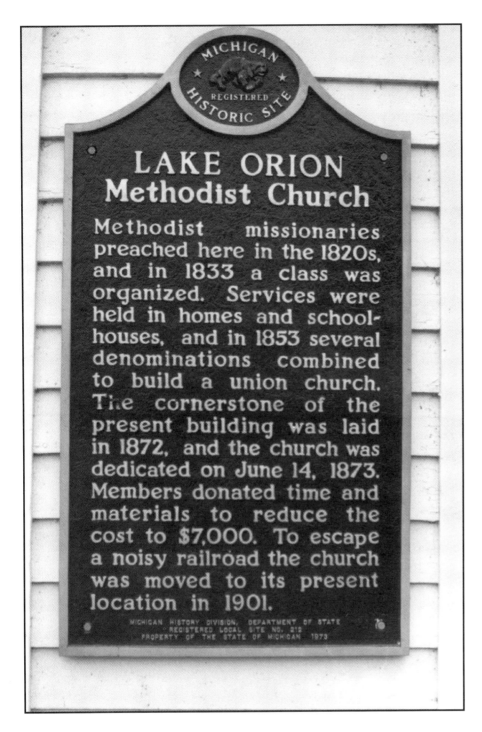

ROLLED CHURCH TO NEW SITE. In 1901, logs were placed under the church. With the help of draft horses, the church was rolled four blocks down the street.

sent a 100-year award that's posted outside the church.

Sleeping In Church

Each time after one of those jabs in the ribs I mentioned earlier, I would tell my wife, "I can't help it, Pam, it's hereditary." I know she doesn't buy that line, but I guess there's some truth about it.

Having gotten up early on Sunday morning to milk and feed the cows, Dad was a chronic church-napper, especially in a hot, stuffy church. But he had an even harder time getting away with it.

Mom always took care of things when she was sitting next to him, but over the years Dad learned to station either Janet or me between him and Mom. But no matter where you'd sit, you couldn't avoid Reverend Herb Hauser.

Rev. Hauser was a good friend of the family and he knew about Dad's inkling to fall asleep from time to time. And he'd stop right in the middle of his sermon to wake his friend up.

When Dad would be dozing off, Reverend Hauser would come up with a bible verse and say, "That was from the book of...J-O-H-N!!!" (He'd shout the "John!").

Because my Dad's name is John, it worked every time, and as we left the church, the Reverend would say, "John, you almost hit the ceiling fans today."

Later on, the condition hit me as well. After working all morning in the barn, spending an hour in a hot, stuffy church was hard to do. On hot days, especially, Mom sat between Dad and I to make sure there was no sleeping. And if it happened, she wasn't shy about getting physical—with the same elbow-to-the-rib cage my wife's made famous at our Wisconsin church.

Shining Shoes

This isn't one of my most intellectual moments, but I guess every kid had days like these. We always got all "dudded up" for Easter Sunday. When I was 11, I found Dad's old shoe-shine kit the night before Easter and decided to polish my shoes.

I spent nearly an hour polishing those shoes with black Kiwi polish. But I went a little overboard. I thought it would really be cool if I polished the soles completely black.

After finishing the job, I left them in the kitchen overnight. After getting dressed in my brown suit and blue clip-on-tie, I came down to put the shoes on and be off for church.

Without giving it a thought, I laced up the shoes and walked out into the living room where everyone was waiting. After my parents gave the usual, "My, you look nice" to every kid who hates getting dressed up, I proudly exclaimed, "Look what else I did"—and showed them the bottoms of the shoes.

I didn't quite get the response I anticipated. Mom's mouth dropped open and Dad sprung up from the chair and ran right past me, glaring at the black tracks covering the green carpet. After some yelling, the shoes came off and I was sent back upstairs to get my old brown "buckle" shoes.

Sunday School Days

Sunday school is a rite of passage nearly everyone must go through at one stage or another. I wasn't a huge fan of Sunday school, partly because of those itchy Sunday School clothes.

When I was about 9, Mom made me wear a pair of wool dress pants that felt like sandpaper on my legs. To relieve the problem, I started wearing my cotton pajamas underneath the pants.

One day during the Bible lesson, some kid tackled me and my pajama leg showed from under the wool pants. Everyone— including the girls—saw I was wearing plaid PJ's. I was really embarrassed.

When I was a kid, Sunday school was from 9:30 to 10:30 a.m. and the main church service was at 11 a.m. After every lesson, I'd walk down to my Grandma and Grandpa Tarp's house in town until my folks were out of church. But every Sunday, I passed the Van Wagoner drugstore in town and bought a farm or sports magazine, drank a chocolate soda and killed an hour or so.

Jesus Threw Up?

One of my parents' favorite "Sunday school" stories involved my sister, Janet. When she was about 4 or 5 and in the pre-school religion classes, we picked her up after one Sunday morning session and Mom asked, "Well, how was Sunday school today?"

A bit confused about the world of religion, Janet answered, "Not too good. Jesus threw up and wasn't there today." After bursts of surprised laughter, we discovered Janet thought Jesus was her Sunday school teacher, who had been out with the flu that day.

A Knockout Punch

My Uncle Sherm and Aunt Helen also were married at our church. "I thought I was going to pass out at my own wedding," Helen says of their wedding 45 years ago.

"During the ceremony, the minister asked us to kneel at the communion rail and told everyone to bow their heads and pray. He put his hands on our necks and cracked our heads right into the rail."

Sherm, who also thought he'd be down for the count, quickly adds with a chuckle, "Maybe it was a sign of what our marriage was going to be like."

Times Are Changing

In the old days, church was as big a part of the community as the village hall and the school system. Things are a lot different from the days when farmers drove their families into town in their horse-drawn carts, tied their teams to hitches in lean-to's or tie-stalls and gathered for picnics after the service.

"Church really played a big role in rural communities," Dad says. "It's changed now that people have cars and can go further."

Church was the center of all social life—especially for farmers in those days. It doesn't seem to be that way today. People don't seem to have the time anymore.

The benefits reached far beyond hearing scripture. Church was an opportunity to get away from the work, sit back and relax with a cup of coffee and find out what your neighbors were really up to.

THE FAMILY BIBLE. This Bible has been passed down among family members for generations.

Christmas Farm Memories

SOME OF the Lessiter family's favorite memories from the family's Centennial Farm are of Christmas.

I am reminded that the beautiful Christmas Story is alive and meaningful only because it was recalled and passed on to us over many years from generation to generation.

The First Tree

Do you remember your first Christmas tree? The glittering tinsel and the shining ornaments? The excitement you experienced as you gazed at the beautiful tree with child-like wonder? Do you remember the breathless anticipation you experienced just looking at the packages piled under the tree?

TREE DECORATING. Mom, Dad and Janet put the final touches on the 1957 Christmas tree. Christmas was always a big occasion for the Lessiter family.

CHRISTMAS ON THE FARM. At 18 months, Mike was already into farm animals in a big way. But the most excitement that year was the large number of Christmas presents that needed to be opened.

Do you recall the tree lights—or were they candles in your earlier days? Were there strings of popcorn or cranberries on the tree?

Too little to understand Christmas itself, I was thrilled by the beauty, the excitement and the joyous spirit prevailing in our home at Christmas.

Do you remember the Christmas you received the most beautiful doll in the world? It was probably just an ordinary doll, yet it may have become your favorite of all time. That doll set you apart as a little mother. It may have caused your eyes to light up and gave you something to really cherish.

Perhaps it was only an old rag doll or a Campbell's Kid doll, like mine. Maybe it had a real China head and a French kid body. Maybe it cried or had real hair. Yet it was special and it has

---

"That old rag doll was special and will always have a special place in my memory..."

---

remained special all these years—tucked away in a special corner of your memory.

Do you remember your first Christmas at school or Sunday School? Perhaps it was a little one-room school or country church. Did you forget your lines for the school or church play in the excitement and stagefright of the moment?

Were you perhaps among that select group chosen to be an angel? Did you wear a cheesecloth robe and a halo? Did you have one of the little high-pitched voices that sang so sweetly?

Do you remember Santa Claus, that jolly, loud-voiced, foot-stomping man whose beard loosened, whose stomach had a way of falling and shaking and whose voice had a familiar ring?

Santa Claus always came in singing "Jingle Bells" to the sound of sleigh bells. He always handed out oranges and candies to all of the "good" little boys and girls.

That Special Memory

There must be one of these Christmas memories that stands out above the others. If you look for it, I am sure you will find it tucked away with your other memories.

Do you remember the first Christmas that really meant the birth of Christ to you? There was a deep feeling that came with it which gave a new meaning to the Christmas Story and Christmas carols that year. Plus, there was a joy to your giving, a sense of humility and thankfulness that made Christmas more meaningful and memorable.

Do you recall your first Christmas with your best beau? Was it a party? A dance? A

CHRISTMAS GIFTS. At left, Janet helps Andy unwrap one of his latest toys. Above, Mom and Dad dig in with three or four special presents from the grandkids.

sleigh ride? Caroling? This was excitement too. There were questions and there were problems. Should I give him a gift— or was that being too presumptuous?

Perhaps a gift of my picture. Should I ask him for Christmas dinner with my family? Or will he want to spend the day with his family, just as I do mine?

Where To Go?

I had always spent Christmas with my family and it just wouldn't seem right not to spend the day with them. But then I would like to spend Christmas with him, too. Who comes first—my family or my beau?

You remember the problems that came with that particular Christmas. Thank you, God, for understanding parents and for their love.

Do you remember your first Christmas away from home? You tried so hard to be brave and not spoil it for others...and really succeeded. But you weren't really so brave. Deep down something was missing.

Do you remember your first Christmas as a parent? Now Christmas had real meaning since you now knew how Mary and Joseph had felt as they looked at their baby.

You searched your memory

"I wonder if the children will always want to come home for Christmas..."

for the Christmas traditions you loved so much so you could pass them along to your children. The fun of everything, the planning, the sharing. Everything took on more meaning. Each year you added to it as you tried to make the Christmas season such a happy one for your children. You wanted to be sure Christmas did not lose its true meaning in all the glamour and excitement.

Cookie Time!

Do you remember your first Christmas as a grandmother? Of course, you wouldn't dream of spoiling your grandchild. I'll guess you got out that old recipe for the decorated Christmas tree cookies which take so long to make, but which your own children liked so much.

Did you wonder if your children would still want to come home for Christmas? And how many times did you say, "Now Grandpa, you know very well that child isn't old enough for a train. But a doll, that would be just right." And, "Did you ever see such a beautiful baby, and such a good one, too?"

Do you remember your first Christmas without the loved one with whom you lived so many years? Try as you may, but it will never be the same without that loving smile, that togetherness and that deep something that was

CHRISTMAS AND THE KIDS. It was always a very special occasion for everyone when the seven grandchildren were all together at Grandma and Grandpa Lessiter's house for Christmas. Here, Debbie helps cousin Andy open a present while Mike sneaks a look at the latest Fisher Price truck.

BUSHELS OF PRESENTS. What I remember most from my early days were Aunt Helen and Uncle Sherm carrying in bushel baskets filled with gifts at Grandpa Tarpening's hosue on Christmas Eve. Here, Aunt Helen helps son Tim Tarpening try on a new jacket while brother Steve plays with a jack-in-the-box.

never lost between the two of you. But you remind yourself he had wanted a happy Christmas and thus you are sustained by your memories.

Well, there they are. All are Christmas memories. All are Christmas experiences. All are marks of growth. And they are beautiful.

Your favorite? Well just as some Christmas days are green ones and some are white ones, they are all lovely and it is hard to decide. I've looked at them all and can't choose. Can you?

A GOOD BOOK. Uncle Neil checks out the latest animal health book which he found under the Christmas tree. With a growing herd of Simmental cows, the book came in very handy many times.

The Old "What-Not" Shelf

ONE OF THE REAL delights of my Mother's childhood when visiting her grandparents farm was seeing the old "what-not" shelf sitting in the front parlor.

Handed down from generation to generation, this 5-foot tall, walnut wood "what-not" now graces the living room of my sister's living room and always delights all of the children who visit the farm.

It has five graduated shelves of varying heights and features many decorative wooden spindles and knobs.

For nearly 50 years, it played a key role in the decor of my Mother and Father's living room. Along with an assortment of special knick-knacks that sat on the five shelves, it served as a real conversation piece for many youngsters and adult visitors as well.

Over the years, Mom replaced a few curios she had viewed on the "what-not" as a child. Yet she could always vividly recall the story behind each one for visiting youngsters, particularly her seven grandchildren.

Neat Little Things

Two little solid brass candlesticks were a gift to my Mother from her older sister, Margaret. They had been purchased in Scotland by Mom's Aunt Kate when she was a Canadian nurse

> *"Mom loved to tell the stories behind every item on the what-not shelf to each of her seven grandchildren..."*

stationed in England during World War I. Two lovely blue magnolia vases were given to my Great Grandmother by one of her sons when he returned from seeking his fortune in western Canada.

The beautiful cranberry glass jar on its silver standard was a wedding gift to my Dad's parents when they were married in 1895. The glass high top shoes were another favorite of Mom's.

So was the little snuff box which came from Scotland when my Great Grandfather came to America—nobody really knew why he brought it since he never used snuff.

Two little salt dishes were among the Willow ware family dishes used by my Great Grandparents. A little cup and saucer was a gift to my Grandfather on his fourth birthday.

The delicate little vase was a treasured gift of my Mother's Aunt Sarah, the oldest daughter in her family. It stood between two hand-painted little pitchers given to my Great Grandmother. The heavy glass paper weights were souvenirs of a trip to Niagara Falls.

A Story For All

Those cute little items found on the five shelves of the "what-

ONE-OF-A-KIND FAMILY TREASURES. This old "what-not" has been handed down from generation to generation in the Lessiter family. The heirlooms kept on its five shelves represent many decades of history for family members.

GRANDPA AND GRANDMA. Mom liked to tease the grandchildren and tell them these dolls were patterned after their grandparents.

CRANBERRY GLASS JAR. Along with a silver standard, this was a wedding gift to Dad's parents when they were married in 1895.

not" each had special memories to Mom. And when one of the grandkids would carefully pick up an item and ask Mom to tell a story about it, she was always ready. She loved to tell why each particular item had been very special to her as a child.

When Hollywood Came Calling

NONE OF THE MEMBERS of our family ended up going to Hollywood for the annual late March Oscar Awards that year, but Dad and Mom did have their 15 seconds of fleeting glory—as movie stars.

I think it was Andy Warhol who originally said it and the saying goes something like this, *"Everyone is entitled to 15 seconds in the limelight sometime during their lifetime."*

While Mom and Dad reaped many honors over the years, this particular one came right down to the 15-second maximum.

Movie Stars Or Not?

Detroit Edison was the electrical utility serving our area of southeastern Michigan. Having worked as an electrician before coming back to the family farm, Dad had met many of the officials running the farm programs for this public utility over the years and also through his work on the Michigan 4-H Council.

He also went on one of the New York City trips with a number of 4-H kids from Michigan sponsored each fall by Detroit Edison. Actually, I had taken this trip a half dozen years earlier as the state 4-H electrical demonstration winner.

So when the utility's communications department was looking for a farmer to feature in a new right-of-way movie, Dad's name came up.

As cooperative as anyone could ever be, Mom and Dad told the movie producer to bring the camera crew out and they'd be willing to "star" in this big screen spectacular.

A couple of weeks later, the camera crew made the 35-mile trip from Detroit, set up their equipment and explained in detail what they wanted Mom and Dad to do. About the time they were ready to shoot, the sun went behind the clouds.

The rest of the morning was spent with everyone sitting around the kitchen table drinking coffee and talking. When

IT WAS JUST A MOVIE on utility line right-of-way, but it turned into a major production when a portion of the filming was done at our Lake Orion farm.

FOREIGN GROUPS, such as these utility engineers from Korea, often visited the family's farm to talk with Dad and to learn things such as how you could get eggs from 3,000 layers without having a rooster.

noon time came, like any good farm wife, Mom wouldn't hear of having the crew go to town to eat dinner.

Lousy Film Weather

The afternoon sunshine was no better and no film was shot that day. The crew finally packed up about 5 p.m. and headed back to Detroit.

The next day, they arrived around 8 a.m. with high hopes of quickly wrapping up the project. Then it started to rain...and it rained...and it rained.

In fact, it rained all day long. Again, the crew sat around the kitchen table, drank coffee, swapped stories and ate another of Mom's delicious dinners before eventually heading home.

The next day was the third day the camera crew had spent at our place. And again it rained all morning.

Finally, after dinner (the third one Mom prepared for the crew), the sun broke through the clouds and the crew shot their needed right-of-way footage and did the lengthy interview on the importance of farmer and utility company cooperation with Dad.

After spending 3 full days at the farm, they thanked Mom and Dad for being super hosts and were on their way. And Dad finally got back to farming!

A Star Is Born?

Well, the film-making was the talk of the town. Especially among the other five couples that were part of the twice-a-month Saturday night bridge group my Mother and Father belonged to for more than 40 years.

The other couples in the bridge club kept bugging Clare Chapin to get a copy of the film so they could see their favorite movie stars in action. Clare managed the Detroit Edison office in Lake Orion and finally secured a copy of the right-of-way film to show to the bridge club members.

One Saturday night between bridge games, the projector rolled and the film came up on the bright white living room wall. The bridge players got their chance to see the result of the camera crew spending 3 full days at Lohill Farm some 18 months earlier.

Maybe you've guessed what actually happened by now. In the entire 27-minute film on right-of-ways, Mom, Dad and the

---

"In the 27 minute movie, our farm scenes made up only 15 seconds..."

---

farm scenes showed up in only 15 seconds of the film!

While the film was showing, the bridge players kept looking for more shots of Mom and Dad. After the lights were turned back on, they said, "That's it? That's what you've been bragging about for 18 months as your chance at Hollywood stardom?"

Mom and Dad never quite lived this personal Hollywood experience down.

Yet there was much more to

THE AD THAT MADE OUR OLD BARN FAMOUS in national snowmobiling magazines is shown at right. Dad rented the barn for a full day's shooting by a Detroit area advertising agency who created this unique ad. They even wiped down all the wood in the old barn with linseed oil to make it shine.

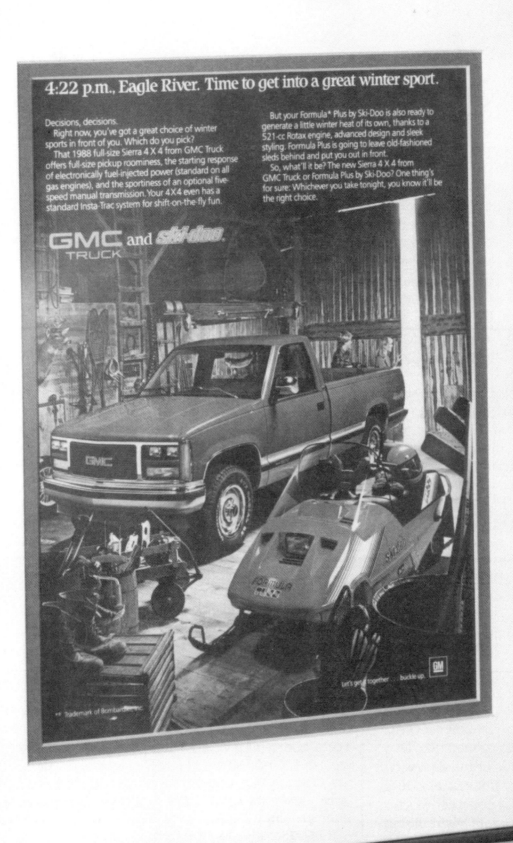

come for the farm when it came to appearing in the national spotlight.

Shooting Ads

Lohill Farm also enjoyed one other film-making experience a few years back. This time it was as a scene for a magazine advertisement.

One day a man stopped at my sister's house on the farm, explained he was scouting sites for a magazine ad and wondered if he could take a look in the big barn. She told him to go ahead.

Later, he called Dad and told him he was an art director with a Bloomfield Hills advertising agency that handled the GMC truck account. He wanted to rent our big barn for a day to shoot some advertising photos.

Dad rented the barn—with one typical farmer's stipulation: There was to be no smoking by any crew members in the barn because of the possibility of starting a fire with all the hay and straw stored there.

I wasn't home when these photos were made, but I've heard many of the details and also have a photograph of the ad sitting behind me on my office wall as I write this.

The photo crew showed up at 8 in the morning and didn't finish shooting until 11:30 that night. They certainly got the full value out of their 1-day rental fee.

An ad that combined a GMC pickup truck with a Ski-Doo snowmobile later ran in several snowmobile magazines that winter. It was shot in early October, but the tiny slit where the big doors were open made it look like it was snowing outside.

When I first picked up one of the snowmobile magazines with the ad on the back cover at the local drugstore, I was really amazed. The old barn had never looked so good.

It was a sharp looking ad with a bright red polished GMC pickup truck sitting in one of the barn's six big bays with a similarly polished bright red Ski-Doo in the foreground.

As Dad later told me, the crew brought its own portable generator, plenty of lamps and reflectors to properly light the barn and lots of props.

Speaking Of Friends...

Living on the farm, there was always one thing which Trista Roberts hated to hear from her friends when they visited.

It was, "Let's go see the animals."

The problem was kids growing up on the farm get to see and feed the animals day after day. Thus it's not any treat to go to the barn again.

"But when my friends came over after school, we always ended up going out to the barn to see the animals," recalls Trista.

She felt there were much more fun things to do. "We had a big rope hanging in the barn that my cousins and I used to swing on," says Trista. "Don't tell my Dad, but we used to pile straw under the upstairs trap doors and jump down into the straw from upstairs.

"I used to drive Mom nuts because I'd climb into the hay lofts when I was too young."

Those props included an old canoe hanging above the old granary area, an antique child's wagon, crates, a 55-gal steel drum filled with hockey sticks, winter boots, a bicycle or two, a

"They shot the ad in October, but they made it look like it was snowing outside..."

deer's head hanging from a hay mow door, snowshoes, an old license plate nailed to the wall, wooden barrels and numerous other items the camera crew found stored in the barn.

The crew spent plenty of time cleaning up the barn for the photos. Both the floor and walls had been swept, then wiped clean with linseed oil. The old barn wood gleamed in the snappy-looking four-color photo.

A young actor and actress completed the photo, looking out through the big barn doors at the falling snow (faked) and wondering when they might be able to pull their snowmobile and pickup truck out for a trek through the nearby woods.

Dad says the crew cleaned up really well after using the barn all day and nobody smoked a cigarette in or near the barn.

That year for Christmas, we had one of the magazine ads framed for Dad. It serves as a wonderful remembrance of one of the days when a film crew came to the farm.

It's certainly a much more permanent momento of the farm's two attempts at Hollywood stardom than the fleeting 15 seconds in the utility company's right-of-way movie.

The Murder At Lohill Farm?

AS YOU CAN probably attest, some very strange things happen from time to time on farms. But I can't remember anything stranger on our farm than the day police officers approached Dad and me in the barn, mentioning the word "murder."

Finding The Clues

Like most kids, farms were fun areas for playing—and our place was no exception. Kids from nearby subdivisions often cut through our farm when walking, and our woods and the farm's own lake were hot-spots for playing. And although we couldn't do too much about it, our 160-acres were also popular motorbike and snowmobile grounds for many neighbors.

The swamps on our land (more commonly known today as wetlands) also attracted a number of kids from nearby homes. Here they'd find frogs, snakes, mice, grasshoppers and other kinds of "wildlife" they couldn't find in their backyards.

One summer day when two boys, about 8- or 9-years-old, were exploring the remote swamp on the far end of the lake, they spotted something that sent chills up their spines. There, in one big pile were a bunch of

THE SWAMP, where the brush is growing tall next to the lake, is where the scared kids found the old bones and dried blood which they thought was the result of a violent murder.

large bones, with patches of dried, red blood still clinging to them!

I was home from college that summer and was bringing the milk cows in when I saw those kids. I waved and started to walk over and say hi to them. When all of a sudden, they spotted me, turned and sprinted toward the back fence and home.

Really Shocked!

Apparently, the terror-stricken grade schoolers ran all the way home and told their parents about the dead body on our land. Living only 35 miles from Detroit, we read about murders in the daily newspaper all the time. These kids never dreamed of murder in the 2,700-population town of Lake Orion, but on our farm that day, murder had stared them in the face.

That afternoon, they repeated and repeated their stories for their parents. They were playing in the fields and found a pile of bones laying near the edge of the swamp. The bones were stacked on top of each other and they even thought maybe there was more than one body there.

Calling All Cars

Dad and I were finishing the milking when I saw a police car roar down our gravel driveway, kicking up dry dust all over the place. Two police officers got out and rushed into our milkhouse, which was attached to the big barn. "Is your Dad here?" I was asked.

"Yes, sir," I answered. "Is something wrong?"

"Listen, kid," said the rattled cop, "just get your Dad!"

I ran and got Dad and we walked over to the policemen standing by the edge of the barn. Dad greeted them, saying, "Hi, officers, what can I help you with?"

I quickly got the idea this was no social visit. "John," one of the police officers said sternly, "we have reason to believe there's a murdered body on your land."

I couldn't believe it. Dad asked what happened and he was told two kids found a body on the back end of the farm. My heart was beating and Dad was just as surprised as I was.

"Let's go take a look," he said, and we got in the back seat of the squad car.

As we drove down the lane toward the swamp, there were people already standing in the field. When we got out next to the spot, we saw two kids in T-

"Okay, show us exactly where you found the body, Peter..."

shirts and shorts standing with their mothers and fathers. The two terrified kids I'd seen earlier stared at the ground, afraid to look at either my father or me.

"Show us where the body is, Peter," the cop said, motioning one of the kids to lead him to the bones. He reluctantly left his father's side and marched slowly into the swamp toward the site.

Following the boy along the edge of the wet, bug-infested path were the two police officers, Dad and me. As my boots squashed along the damp ground, I didn't have any idea what to expect and the tension was building.

When Peter stopped and pointed and we grew closer, Dad took a look and let out a bellowing laugh.

The Real Story

Unfortunately, every now and then a young calf died on our farm. Since it didn't make any sense to pay the rendering company to pick up a 90-lb. calf, we usually buried them ourselves.

But when they died in the winter, we rarely tried to bury them in 15-degree below zero weather. Occasionally, Dad would have our hired hand drop the corpse back in the woods where the bones wouldn't damage our tractors or tillage machinery.

Apparently, our hired hand dropped this calf out by the swamp. And when spring came around, he forgot about going back and burying it. The bones had probably sat there for a couple of years before the boys stumbled across them.

A Bit Embarrassed

When Dad was done laughing, he told everyone there was nothing to worry about. "That's just a dead calf we couldn't bury a couple of winters ago," Dad said. "The ground was frozen and the hired hand just tossed it out here."

The police officers were a bit embarrassed, but they hadn't had much experience with murders. The kids and their parents both apologized and before long the kids, who were scared to death that afternoon, were back playing on our farm again.

I'll admit I was pretty scared too as we made the drive back to the swamp. But I remember working the rest of that day with a big, big grin on my face.

1957 And 1958 ...B.D. And A.D.

ONE YEAR in our farm lives, our family seemed to date every event as B.D. and A.D.—before Dietrich and after Dietrich.

That was the first name of our exchange student from Buel, Germany, (located directly across the Rhine River from Bonn, which was then Germany's capital) who became a member of our farm family for a full year.

Another Senior!

Dietrich Ristow was a very interesting and very likable addition to the family's household. He arrived within just a few weeks after I had left home to become a freshman at Michigan State University. For the second straight year, Mom, Dad and my little sister once again had a high school senior in the Lessiter household.

With Dietrich there, everyone broadened each other's knowledge and we shared our different ways of living and ideas, routinely comparing our priorities, philosophies and differences between U.S. and European living.

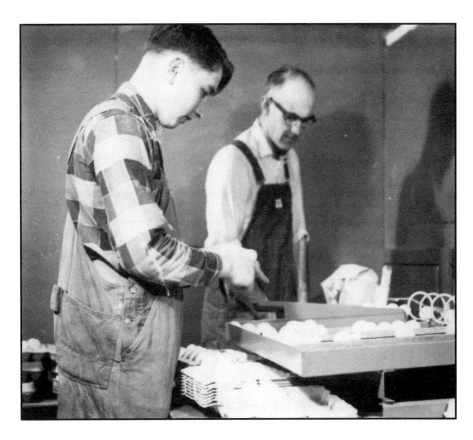

GRADING EGGS. During his exchange student year, part of Dietrich's part-time farm job was to help Dad grade eggs each day from 3,000-plus layers.

PITCHING IN. Dietrich and Frank take a Saturday morning crack at cleaning out the heifer barn.

As Mom used to say, what at first seemed to be a frugality bordering on stinginess was more easily understood as a feeling of insecurity and uncertainty. This was based partly on the fact that we realized Dietrich had experienced having his home bombed twice and his family having lost almost everything a decade earlier during World War II.

Prior to Dietrich's arrival in September, I spent the month of July living in his home in Germany as an exchange student. This helped me get to know and enjoy his family.

The correspondence which Dietrich's mother and my Mom later carried on for a number of years was a delightful memory my own mother always treasured.

Even as we learned a greater tolerance for others and their way of life, our family received

> *"Since Dietrich had a younger sister at home, this was no new encounter for him..."*

a much greater appreciation and respect for our own country as seen through the eyes of Dietrich. Even our command of the English language improved and became more precise as everyone helped him improve his English—especially my English teacher Mother.

Big On Education

Our family and Dietrich's new-found high school friends soon learned to share the emphasis which he and his family placed on education, along with many other aspects of our life which we have so long taken for granted.

Mom and Dad quickly came to think of Dietrich as their own son and took special parental pride in his many American accomplishments. They took real pleasure in his solution to the dilemma of what to call them. "Mother and Dad" soon became a real compliment to both of them.

When Dietrich and my younger sister, Janet, frequently disagreed, everyone knew a real

FRENCH HORN PLAYER. With several years of experience in Germany playing the fluegel horn, Dietrich quickly caught on to the American counterpart, the french horn. During his senior year, he was an active member of the Lake Orion High School band.

He also played french horn in the high school band and was part of the marching band that played at football games. He'd played the "fluegel horn" back home in Germany, so adapting to the American counterpart was no problem.

His interests included photography, bird watching, reading and listening to music. While everyone tried, nobody ever truly interested him in farming. But he pitched in with the farm chores and earned an allowance for farm work just like I had done.

He definitely had a scientific

> *"Mom used to hope this 'second son' would someday return with his family for a visit..."*

mind and later became a nuclear physicist in Germany.

Like most teenagers, Mom found Dietrich could be difficult at times. Yet everyone respected his differences of opinions and his intense loyalty to his homeland which he expressed upon

brother and sister relationship had been established. Janet occasionally cut him down to size (as she could do with me—her very own brother!) with no danger of damage to international relationships.

Since Dietrich had a younger sister at home, this was no new encounter for him.

His special "Grandpa" relationship with my grandfather was pleasing to both of them and Mom. Dietrich would occasionally stay overnight with "Grandpa Tarp" in Lake Orion.

6 Years of Physics?

Mom and Dad were very happy to see him accepted by the high school student body and by our family's friends. Everyone took great pride in his graduation as an all-A student. While some classes were harder than others for him, his first year physics class was a breeze since he'd already studied the subject for 6 years back home in Germany.

Dietrich was also a loyal supporter of the school's athletic teams and gave track a shot until painful shin splints did him in.

Nobody in the school or town could match him in ping-pong and he introduced the soft padded paddle to local players.

PHYSICS WAS A PIECE OF CAKE. Dietrich is shown here with Mom, Janet and Dad. The Lake Orion High School counselors enrolled Dietrich in high school physics class, even though he had already studied the subject for six years back in Germany's schools. Returning to Germany and university studies, Dietrich later became a successful nuclear physicist.

occasion when he felt it suffered too much by comparison.

Even so, he learned respect for our judgments and the few rules my parents imposed on him.

Big Gains From Visit

So it was with a little amusement—but great interest— that the family watched his increasing attachment to many of the luxuries we enjoyed and took for granted. At first, he had seemed to feel these were often signs of softness or even laziness.

Riding when someone could walk almost seemed inexcusable to Dietrich. Taking a daily shower or bath seemed like a waste of water.

He was constantly amazed at the generosity and hospitality of our friends and American people in general. He liked our family's social life and friends. They, in turn, liked him.

No Easy Goodbyes

When it was time for Dietrich to go home to Germany, it was with genuine regret that everyone said goodbye to him. He had made himself a part of our lives and we had all come to love him.

Mom used to hope "this second son" would someday return along with his family for another special visit.

It didn't happen during her lifetime, but my sister did get the chance to visit Dietrich and his family while she and her hus-

> *"He had made himself a part of our lives and we had all come to love him..."*

band, Neil, a U.S. Army baseball player, were stationed in Germany.

"Because of Dietrich, our own son and other exchange students," my mother once wrote, "I am sure our country's foreign relations have improved. Certainly our understanding of one another is much better as a result."

TOUGH PLAYER TO BEAT. Using a unique, yet definitiely legal, sponge rubber paddle which he had brought with him from Germany, practically nobody in all of Lake Orion could ever beat Dietrich in a ping pong match.

Corn Was King

WITH PLENTY of livestock to feed, corn was the most important crop grown on the farm. It was rare that any corn was sold as the crop was effectively marketed through the farm's livestock.

There were also years when we ended up buying more corn from other farmers or the mill in Oxford to feed our livestock. Yet it was a good year when we had enough corn to finish out the year until the next crop was ready to be harvested.

For years, the farm's crop rotation called for one year of corn, one year of small grains (wheat, oats, barley or sometimes rye) and two or three years of interseeded alfalfa.

Following corn silage harvest, wheat would be seeded in the fall. If we were planting a spring crop, such as oats, barley or rye, on ground where corn had been harvested for grain, alfalfa would be seeded in the spring at the same time as small grains.

For a number of years, we also grew a few acres of potatoes.

Silos located at each end of the big dairy barn were filled every fall with lush, green corn silage. In those days, most farmers followed the Cooperative Extension Service recommendations which called for cutting corn for silage while the crop was still green. Farmers did not let it mature in the field as is done today in order to increase the grain content of the enclosed crop.

The goal was to get your silos filled when the crop was supposedly at its full nutritional value, definitely before the corn was hit by a killing frost. Later research proved this theory was all wet.

In those days, it was the sign of a good farmer to have silage

CORN AND COWS. With a Holstein dairy herd, corn silage and shelled corn were ration mainstays. Along with alfalfa, corn was the farm's major crop.

CORN TOOLS. Binder used to cut corn bundles (shown shocked below), silo filler and hand-powered corn sheller.

juice seeping out between the silo staves or wooden boards as it meant you'd gotten the crop harvested at just the right maturity.

Corn to be used as livestock feed on the farm was cut with a corn binder and the heavy corn bundles were shocked for drying in the field. After field drying was completed in late October or early November, the heavy bundles would be loaded onto a horse-drawn wagon and hauled to the corn shredder where ears would be pulled off the stalks.

The resulting ear corn would be shoveled into a corn crib for further drying. When needed as livestock feed, ears of corn would be run through a sheller and later be ground.

A strong believer in soil conservation, Dad tried the wheel track corn planting system. The minimum tillage system was developed by Michigan State University soil scientist Ray Cook in the early 1950s.

This meant planting corn with a two-row planter into plowed ground without any tillage. But Dad gave it up because it was too rough on the machinery.

Besides planting and harvesting, the corn crop was normally cultivated twice early in the growing season. Later on, herbicides replaced one cultivation.

Handling corn this way was exhausting work. When the combine came into the picture and shelled the corn in the field, a tremendous amount of labor was lifted from the shoulders of farmers who relied on corn as their number one crop.

But as was the case on many Midwestern farms, corn was always king at Lohill Farm.

BEAUTIFUL FALL SCENE. Heavy shocks of corn used to dot the rural countryside at harvest.

Lake Orion's Boys Of Summer

IT WAS FORMER New York Yankees catcher Joe McCarthy who said, *"Give a boy a bat and a ball and a place to play and you'll have a good citizen."*

There's some real truth to those words describing our nation's love of the leisurely game that holds so many dreams and memories for most of us.

When you've got a bag of peanuts, a scorecard and 9 innings in the sun, there's no one experience that better represents America.

If you travel through any farm community after the final patch of snow melts away, you'll often find a group of kids out in the pastures with their bats, balls and mitts, imitating their favorite star's batting stance, making diving catches and sliding head first into home.

Baseball was no stranger to my hometown, either. In fact, our town's baseball tradition extends even further than those games played in makeshift pasture ballparks. Lake Orion was the hot spot for the game's biggest heroes when they needed a break from their 154-game season.

Relaxing On The Lake

Dad remembers when baseball's top major league stars used to visit Lake Orion on the weekends. "In those days, they

"They didn't play on Sunday, so the major league players would come out to Lake Orion after the Saturday games..."

didn't play ball on Sundays," Dad says. "So many of the players would rent weekend cottages on Lake Orion."

After playing Saturday afternoon games in Detroit, players would jump on the Motor City streetcars which stopped at Lake Orion, Flint and Bay City. Since the cottages weren't accessible by road, the players would get off the streetcar and be carried by boat to their respective cottages. Even groceries were delivered by boat in those days.

"They'd all come out, relax and have a great old time," Dad remembers. "They'd fish, drink and play cards."

The Babe's Here!

The New York Yankees' famous Babe Ruth was one of the many ballplayers who regularly visited the calm waters of Lake Orion for a number of years. He probably ate one of his notorious breakfasts at one of the lakeside diners in town, consisting of an 18-egg omelet with 3 slices of ham, 6 slices of toast and 2 bottles of beer.

Ruth, who once said, "I swing

big, with everything I've got. I hit big or I miss big. I like to live as big as I can," sure meant it when it came to eating. Local restauranteurs always had a huge smile when totaling Ruth's check.

When Dad was growing up, Detroit's Ty Cobb and Ruth were the big names of the era, arguably the best players to ever play the game. Cobb was the fiercest competitor baseball ever knew, and probably the meanest player as well, both on and off the field.

He was known for sliding into opponents with his metal spikes pointing straight up and also on occasion would jump up into the stands and beat hecklers until the police were called to break up the fights. He was the dominant hitter, base-stealer and competitor of the era.

The first local boy to make it big that Dad remembers was Charlie Gehringer from nearby Fowlerville. Gehringer became a six-time all-star with the Tigers from 1924 to 1942 and still holds one of the first three spots in seven Tiger offensive records.

Another player, Mickey Cochrane, also made it big with the Tigers. Cochrane came from a well-established farming family in Milford, Mich.

As far as playing, Dad remembers marking lines out in the pasture and playing games with friends after school. When there weren't enough kids around to play a full game, they'd play 'scrub' baseball—the poor man's version of the game which could be played with as few as three players.

Dad was always too busy to make it to Tiger games, but attended a number of high school games and home talent league games in which neighboring towns would play each other.

"When I was in high school, one or two of my friends owned a set of radio headphones," he said. "I always liked to listen to the Tiger games."

Dad remembers an interesting fellow in the area named Peter Milner, who always wanted to play baseball but couldn't run because one leg was shorter than the other. But even though he never would be able to play, he found his own way to stay close to the game.

"He used to umpire sandlot games in the area and would umpire hundreds of games a summer in Oxford and Lake Orion," Dad says. "He was a top-notch ump."

A League Of Her Own

Incidentally, Mom was a pretty fair softball player, too. While she was in high school and home from college for summers, she played on a softball team

---

"Your Mom was a pitcher and a pretty darn good one at that..."

---

representing Lake Orion which played teams from other area towns. "She was a pitcher and a pretty darn good one at that," says Dad.

After my mother died in 1982, Dad and several friends who were also widowers used to get together and occasionally go downtown to catch some Tiger games. I found out Dad had never been to Tiger games before, but really enjoyed them. I guess the sport really can help heal all wounds.

"Lohill Field"

The extent of my baseball "career" was when I owned a Red Rolfe baseball mitt (named after the Tiger manager and former big league ballplayer) and played in the local Cub Scout league.

Like my father did years before, I spent many afternoons playing baseball on a diamond laid out in our pasture. After finishing chores, a few friends and I would play until dinnertime when there'd always be a big pitcher of ice cold lemonade waiting for us.

When I was in seventh grade, a friend of mine and I were the official bat boys one year for the high school team.

Growing up, my favorite Tiger stars were pitcher Hal Newhouser, outfielders Vic Wertz, Johnny Groth and Al Kaline, and infielders George Kell, Harvey Kuehn, Jerry Priddy and Ray Boone.

I was also a big fan of Cleveland Indians pitcher Bob Feller, a farm boy from Iowa who developed his great fastball doing heavy chores on the family farm and pitching into a backstop of 2 by 4's and chicken wire made by his father.

When I was 8 or 9, I used the lawnmower to carve a diamond

THE BABE. New York Yankees star Babe Ruth was a frequent Lake Orion visitor during the summer months. He'd come out to lake cottages with other ballplayers or friends when the team was in Detroit to play the dreaded Tigers.

HISTORY IN THE MAKING. Grandpa Tarp and 12-year-old grandson Frank were sitting in the 1941 Ford listening to the Detroit Tigers play the St. Louis Browns on Sunday afternoon, August 19, 1951. They heard WWJ radio broadcaster Ty Tyson call the game when Bill Veeck sent 3-foot, 7-inch Eddie Gaedel up to pinch hit. While laughing, Tiger pitcher Bob Cain tossed four consecutive balls to Gaedel who wore jersey number 1/8. Tigers catcher Bob Swift was on his knees to catch a pitch still over the midget's head.

in front of the big barn and all by myself "played" every World Series game imaginable. I'd play the ball off the barn, throw pop-ups to myself—even run the bases, dreaming I was Bob Feller or George Kell.

Occasionally, Uncle Sherm and Grandpa Tarp took me to Tiger games at Briggs Stadium (as it was then called) where I saw my heroes play for real.

Front Row Seats

Grandpa Tarp was a true blue Tiger fan and used to listen to every Tiger game on the radio. In earlier days, Tiger radio announcer Ty Tyson didn't travel to road games but would broadcast the game to Tiger fans from the Detroit studio while getting game updates over the telegraph.

Complete with the sounds of a crack of the bat, fans cheering wildly and a husky umpire bellowing, "steeeeeerike!," no one could tell the difference.

I remember one Sunday afternoon when Grandpa Tarp was at our house for dinner when I was 11-years-old. After dinner, we went out to the driveway, sat in the front seats of his old 1941 Ford and listened to the Tiger game on the radio.

That was the day when the Tigers were playing the St. Louis Browns and eccentric Browns owner Bill Veeck sent a 3-ft., 7-in. midget named Eddie Gaedel to bat wearing the uniform number "1/8."

Detroit pitcher Bob Cain was laughing too hard to pitch to kneeling catcher Bob Swift and walked Gaedel on four straight pitches.

Over 40 years later, my son and I saw Gaedel's uniform along with memories of my other favorite Tiger stars at the Hall of Fame in Cooperstown, N.Y.

Our Second Language

No matter where I travel for farm meetings or field days during the summer, baseball is always a universal topic of discussion. Maybe it's because the game is so appealing to country folks, because it's relaxing or because they listen to games over the radio while driving tractors or combines across the fields.

Or maybe it's because the game always seems to bring out the kid in all of us, recalling those memories of "when I hit that big home run" or "when I struck out the town's slugging king."

You can have your fancy ballparks and artificial grass. My fondest memories of our national pastime were playing at Lohill Farm in a self-made field with a few friends in our very own "field of dreams."

Our Role In Rural Newspaper Delivery

AS A FARM JOURNALIST for more than 30 years, I've always been proud of how the Lessiter family played an important role in bringing about rural newspaper delivery in the Detroit area and maybe the United States.

Years ago, people living in rural areas received their newspapers by mail. But if you chose instead to subscribe to the evening newspaper, then you wouldn't receive the paper until it was mailed the next day—when the news was already old.

The publisher and well-to-do

LONG-TIME FARM MANAGER. Sidney Smith managed the 3,000-acre Wildwood Farms for 36 years and is shown in front of the W.E. Scripps mansion on the farm. Scripps purchased a "package" from the U.L. Clark herd in 1916 that included 16 Angus cows, a bull and Sidney Smith as his new farm manager.

HERD HEADQUARTERS. The barn at above right housed the Aberdeen-Angus show herd while the barn on the left was home for the well-known Guernsey dairy herd. The photo at right shows these barns later converted into the Olde World Canterbury Village complex.

son of the founder of the *Detroit Daily News,* William E. Scripps and his wife Nina, started buying land from several property owners in 1916 under the name of Wildwood Farms.

Scripps started with only 600-acres which he purchased from

What Was Wildwood Farms?

A fountainhead of the Aberdeen-Angus breed in North America, the starting place of the Bardoliers, home of International champions and blue ribbon winners—that was Wildwood Farms.

Few herds have had a bigger part in advancing the breed than did the W.E. Scripps breeding establishment at Lake Orion, Michigan.

For more than three decades, it was an important cog in the business through two World Wars, hard times and numerous health problems.

Every famous herd is a combination of top ownership and management—plus good cattle. During its nearly four decades, Wildwood was under one ownership and one management. Its owner, William E. Scripps, national known newspaper publisher, became interested in Aberdeen-Angus many years ago. A relative was running commercial doddies on his ranch in California and the Michigan publisher was very much interested in their uniformity and performance.

When he decided to put Aberdeen-Angus on his 3,000-acre Wildwood Farms north of Detroit, he selected well as to foundation material and managerial material. His manager of 1916, as well as for four decades, was Sidney Smith, who got his Aberdeen-Angus apprenticeship in the pioneer herds of Charles Escher, Judge John S. Goodwin and U.L. Clark. From the Clark herd, Scripps got 16 cows, a bull and Smith, who had helped select the original Clark purchases.

Wildwood, through the years, was the home of many of the famous cattle of the Angus breed.

—*May, 1949, Aberdeen-Angus Journal*

THE BULL PEN. For a number of years a restaurant by this name operated in the old dairy barn.

total takeover from all the land that ranged from Clarkston Road on the north to Baldwin Road on the west to Waldon Road on the the Haddrill family in May and September of 1916. By the late 1920s, he had added farms from Benedick, Dickman, Newman, Lomerson, Cole and other families. Two holdouts denied him

PROUD MAMAS. A good looking group of Angus cows stands in front of the Angus beef cattle show barn at Wildwood Farms.

south to Josyln Road on the east. Jay Gingell refused to sell 89 acres on Baldwin Road and Charles Cole refused to sell 40 acres along Clarkston Road.

By the late 1920s, Scripps owned a 3,000-acre farm which was truly a one-of-a-kind estate in many respects. Living only a few miles from our family farm, Scripps soon became friends with my grandparents.

Grandpa Frank was township supervisor and got to know the new area land baron quite well while dealing with him on local government business.

"Daddy Warbucks"

Scripps was well known for his generosity around Lake Orion. He offered to build a new schoolhouse for area children, although it didn't turn out as planned.

"Some of the area families put tax money into the local school district to build the new school along with considerable money from Scripps," recalls Dad. "But they later objected when they realized their kids would have to walk to school.

"Scripps was disgusted and gave the school district all its money back and put in all of the money himself for the school. The other people then started Carpenter School so their kids would not have so far to walk."

Scripps also donated livestock for charity ox roasts. He also used to set up a movie screen in

"The entrance to each room featured a carved stone panel which indicated what the room was used for..."

his big barn and show free movies weekly during the summer.

13 Farm Buildings

In 1923, Scripps moved the big dairy barn a half mile southeast so he could build the fabulous Moulton Manor which was modeled after an English tudor style estate. The house included many rooms and the entrance to each room featured a carved

FABULOUS MOULTON MANOR. The big dairy barn (shown on page 256) was moved one-half mile south so W.E. Scripps could build this fabulous English tudor style country home in 1923 as his residence on the farm. It later was The Guest House, a rehab center for alcoholic priests.

BIG ON EDUCATION. W.E. Scripps donated a site and most of the money needed to construct this school to educate the children of farm workers and others in the neighborhood. Yet residents objected when they realized their kids would have to walk a considerable distance to school and started what became known as Carpenter School. In later years, this old Scripps School became a church.

stone panel which indicated what the room was actually used for. A carryover from the days when the majority of the population was illiterate, pictures were used instead of words to identify the rooms. Above the music room, for example, you could find instruments and sheet music carved into the stone.

Many years later after Scripp's death, the house became a nationally known rehabilitation center for alcoholic priests.

Few farm buildings feature outstanding architecture, but the ones at Wildwood Farms did. By

FARM MANAGER'S HOUSE. This attractive house served as the home for Sidney Smith and his wife for more than 40 years. The couple were good friends of the Lessiter family and they often worked together on farm issues.

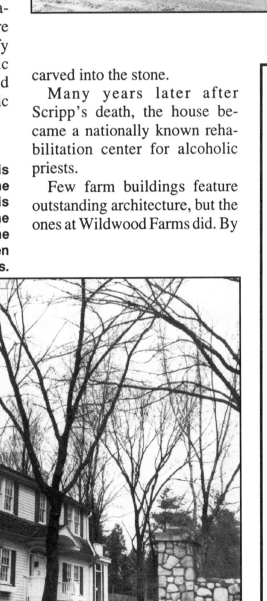

THE ELLIS ISLAND SAGA...

Sidney Smith served as the Wildwood Farms farm manager from its founding in 1916.

One time when Smith and his family were returning by luxury liner from buying Angus bulls in Scotland, one of his sons became sick on the ship.

As a result, the family found themselves unloaded at Ellis Island in New York City's harbor and housed with thousands of immigrants waiting entry into the U.S.

But their Ellis Island stay was short. Sidney Smith immediately called Scripps who placed a telephone call to influential friends in Washington who passed the word to the Immigration Service.

As a result, the Smith family was on their way west via New York Central pullman railway car to Detroit only a few hours later.

GOOD FARM HOUSING. Two-story homes that were modern for the times were constructed all in a row for Wildwood Farms employees. With excellent housing and working conditions, labor turnover on the farm was minimal. These homes later became speciality shops at the new shopping complex.

the late 1960s after Scripps death, much of the farm had been sold and became a large-scale housing complex called Keatington. The barns were rented to several folks who operated shops and in the mid-1990s, the complex was turned into what is now known as Olde World Canterbury Village.

Filled with shops throughout the 13 old barns and tenant houses, the complex offers a special year-round Christmas shopping experience. Other parts of the farm became a county park.

By 1929, Scripps had one of the first farms in the country to be fully mechanized...with hay

HERDSMAN'S HOME. This attractive little house served as the home for one of the farm's numerous herdsmen who took care of the nationally-known Guernsey dairy cattle, Angus beef cattle, Belgian horses, Duroc hogs and Shropshire sheep. The best of all of these breeds were regularly shipped off in the farm's fancy livestock palace railroad cars to shows from coast to coast in both the United States and Canada and always came home with plenty of blue ribbons.

baler, threshing machine, corn sheller and many other labor-saving machines.

For years, the farm was known for its world-renowned blue ribbon winning Angus beef cattle, including many imported from Scotland. Showing steers was always a prominent part of the farm's cattle program. In fact, it was this Michigan herd that showed the first Angus steer to end the long Hereford monopoly at the American Royal Livestock Show in Kansas City, Mo.

One of the great bulls of the Angus breed was raised here by the name of Black Bardolier. This great bull of the breed missed being a show ring steer only because the Wildwood beef herdsman at the time refused to castrate so promising a calf.

The farm also had an outstanding high producing Guernsey dairy herd from which milk was processed in the farm's creamery into butter and other dairy products.

The farm was also know for its Belgian horses, Shropshire sheep and Duroc hogs. The best of these breeds were shipped each summer in fancy livestock palace rail cars to livestock shows around the U.S. and Canada where they enjoyed tremendous success in the show ring.

During the Great Depression, Scripps helped unemployed Detroit workers and their families by donating 28 head of pedigreed Angus cattle, 190 sheep and 1,800-bushels of potatoes to a hot meal program sponsored by city firemen.

He brought a lot of excitement to our sleepy little town and even had an airplane he liked to pilot himself from the farm's private airport. After a couple of near accidents, he relied on his own pilot and later bought an autogyro, a forerunner of the helicopter which he often flew daily from his home to the newspaper offices in downtown Detroit.

Nina Scripps was an accomplished organist who invited musicians from all over the world to give recitals on the pipe organ that was part of their palatial home. With the help of farmhands and area workers, she threw a variety of social functions, hosting many of Detroit's notable businessmen.

One time in the 1930s for example, famed woman pilot Amelia Earhart visited the farm.

The Two-Paper Route

One evening while Scripps was visiting my grandparents' house, he spotted a copy of the *Detroit Free Press,* his chief newspaper competition. Surprised, he asked my grandfather why he read the morning *Detroit Free Press* instead of Scripps' very own evening *Detroit News,* which his father had founded in 1873.

"Your Grandfather explained that by taking the morning *Free Press* we could get the newspaper in the mail the same day it was published," Dad recalls. "If you subscribed to the evening *Detroit News,* you didn't get the newspaper until the next day and the news was old by then."

From then on, Scripps said he would see that we received the *Detroit News* the same afternoon it was published. Scripps already had a driver rush a copy of the first edition of the paper to his home every afternoon, which was 35 miles north of Detroit.

So after talking to my grandfather, Scripps had the driver continue driving another 3 miles to our farm house and deliver the paper seven times a week.

Soon after, officials at the paper pondered the thought of charging for same-day newspaper rural delivery. As a result, they added other rural homes to the driver's route and this became the start of the first rural delivery of any daily newspaper.

"My parents subscribed to the *Detroit News* through Mr. Scripps' driver for the rest of their lives, even after his death in 1952," Dad recalls. "I guess we got a personal daily copy of the paper this way for more than 40 years."

Living in five cities, many daily newspapers have landed on my front porch over the years. But without doubt, the *Detroit News* has always been my favorite paper to sit back and page through.

Maybe it's because I grew up reading it or because it's always had an outstanding reputation.

Or maybe I'm so fond of it because of the relationship of our family and Scripps, which led to a change in the way farmers received the latest news.

BEAUTIFUL SCENES. There were always plenty of beautiful scenic views to be seen on the 3,000 plus acre Wildwood Farms.

"Hi-Yo Silver—Awaaaayyyy!"

TO MOST OTHER KIDS when I was growing up, the Lone Ranger only existed on the radio and in comic books, movies and television shows.

But for me, the Lone Ranger and his trusty horse, Silver, were much more than just characters to dream about out in the wild, wild West.

That Masked Man?

Before I share my memories of the great Masked Man, let me give you some background on the role this 22-year radio series hero played in my life.

It may seem crazy to you, but I want to start by telling you about the two ways we could drive to the Oxford Cooperative Elevator located 8 miles away

THE LONE RANGER. The long-time radio voice of the Lone Ranger lived only a few miles up the road from the Lessiter farm.

where we had feed ground once a week for our dairy herd:

★ One way was to make the normal 5 1/2-mile drive on the blacktopped roads into the town of Lake Orion, located 35 miles north of Detroit, Mich. Then we turned north for 3 miles on Michigan's M-24 state highway to reach the feed mill on the northern edge of Oxford.

★ The other route was to drive 2 miles on the asphalt road, then turn north onto a series of

> *He was known coast to coast as the Lone Ranger for some 14 years..."*

three bumpy, rutty, dusty roads that led us into the village of Oxford and along tree-lined blacktopped streets to the mill.

I always preferred the bumpy, rutty, dusty roads. While we didn't take that route too often, especially in early spring when these roads were often a muddy mess, it was always a treat to take that back road to the feed mill with a pickup truck loaded with ear corn.

Real Country Adventure

It was always an adventure...and I guess that's why I liked the dirt road journey so much. The roads had terribly deep ruts during the spring, were dusty in mid-summer, were covered with leaves in the fall and were lined with towering snowbanks during the winter months.

What I really liked about making this trip was there always seemed to be a fresh approach to life along those country roads.

It was much like driving through a beautiful forest in the summer. We would drive past two beautiful lakes, go past a summer camp for city kids and drive across the railroad tracks.

We would go by several farms with beautiful, attractive buildings. We went by a farm that later became a home for the Catholic Church's Order of Dominican Sisters and still later a business college when suburbia crept into the area in a big way.

Best of all, this backroads journey took us by the home of the long-time voice of the Lone Ranger, Brace Beemer. He was known from coast to coast as the Lone Ranger for 14 years.

The Lone Ranger, who rode that famous white horse, Silver, certainly never let me down when we drove by his big white house along one of these dusty old roads.

Still active as the voice of that famous Masked Man, Beemer lived his role to the hilt...with a big corral out on the front lawn.

Most of the time, I remember seeing a big white horse who answered to the name Silver galloping around the corral. It was all part of Beemer's Lone Ranger image.

I never saw Brace Beemer

LONE RANGER AND SILVER. Clayton Moore played the Lone Ranger on television, but a portion of the Masked Man's history actually centers around the area near Lohill Farm.

OUR OWN SILVER AND SCOUT...

While I never saw Tonto or his horse Scout during these backroad journeys to Oxford, we had a team of horses in our barn with the same names as the trusty steeds which Tonto and the Lone Ranger rode to stardom.

When I was a first-grader, a neighboring farm owned a pair of Belgian horses named Silver and Scout. Because these horses had at least a namesake tie to my favorite radio show, I begged my Dad to buy them. Later when he needed an extra team and purchased them, no one was prouder of these two horses than I.

Actually, Silver and Scout were a hard-working pair of horses until old age caught up with them. One major problem was Scout had a tough time with allergies. In later years, when we used these Belgians to rake hay, Scout had to stop frequently and spend 5 minutes catching his breath every 100 yards or so.

Finally, the veterinarian said he was pretty sure the old horse was actually allergic to hay—something he'd certainly eaten and been around all his life.

I remember well the day when we took a saw to the wooden tongue on the hay rake and cut it in half so we could rake hay with the tractor. Pretty soon, Silver and Scout were gone to a new home on another farm.

around the house or corral, but I sure relived the Lone Ranger radio series every time we drove by that house.

How It All Began

Taking over a radio station that was dropping hundreds of dollars a week during the depression, a new manager at WXYZ in Detroit wanted to develop a new show, a Western of unprecedented wholesomeness with a bigger-than-life hero distinguishable from all others.

The result was the Lone Ranger, his faithful Indian companion, Tonto, and the Masked Man's well-known cry, "Hi-Yo, Silver!"

The show's producers gave the Lone Ranger a strong code of ethics, good grammar, grim dedication, grit and much more. It was their belief that kids would buy it. And they did, right from the start!

For the next 22 years, the famous Masked Man and Tonto filled the air with plenty of Western thrills. "A fiery horse with the speed of light, a cloud of dust and a hearty Hi-Yo Silver! The Lone Ranger rides again!"

When the show was less than 4-months-old on the Detroit radio station in mid-1933, it drew over 24,000 replies to its first premium offer.

A remarkable 70,000 children turned out to see the Masked Man at a Detroit area school function, creating pandemonium for police and causing the producers to later keep the Masked Man under tight wraps. Even so, the show's fame grew rapidly.

Several men handled the voice of the Lone Ranger in the early days before the radio station staff decided on a permanent person.

Earle Graser played the role for 6 years until he was killed early one morning in 1941 just outside Farmington, Ill.

His death created a crisis. For a time, the Lone Ranger was actually written out of the show, leaving Tonto to carry on alone.

A New Voice!

When the Lone Ranger did come back, it was with the slightly huskier tones of Brace Beemer who had earlier worked as an announcer on the 30-minute radio show. Beemer served

"I'll always remember the big white horse standing proudly in the corral..."

as the voice of the Lone Ranger from 1941 through May 27, 1955, when the show finally went off the air after 22 years.

A farmer's son, Beemer had won a Purple Heart for service in World War I and like more than 50 WXYZ employees, he had earlier worked on the series from time to time.

Later the show, with different actors, enjoyed a run of 8 years on ABC prime time television. For years, it held down the extremely valuable 7:30 p.m. Eastern Standard Time prime spot on Thursday nights from coast to coast and ended up as the 69th longest running television series in history.

Its 221 episodes ran in prime time from the 1950 through 1957 seasons and in daylight times after that. Even today, the show can be seen in syndication on many local television stations. And many of the original Lone Ranger radio broadcasts are still available on tape.

So that's the story of the special impact the Lone Ranger had on me when I was growing up.

Those Country Roads

What else did I see as a kid along these country roads?

Well, there was a little spring by the side of the road at one point. There were many beautiful fully grown trees.

Best of all, there was always something different along the road we traveled on the way to the feed mill. That's what always made this backroads country journey extra special.

There was none of the hustle and bustle of busy highways on this journey. We often took our time journeying along these back roads because we really didn't have any choice since the bumps practically bounced us right out of the pickup truck seat.

When we were running late, we often had to hang on for dear life or slow down since this was before anyone had ever required seatbelts in a pickup truck.

That's all changed now as these backroads are blacktopped and there are hundreds of suburban homes along the way.

Hi-Yo, Silver!

But I certainly remember these dusty country roads and scenes very well from when I was a kid. And I'll always remember the Lone Ranger's big white horse standing proudly in front of Brace Beemer's home.

The Lone Ranger was more than just a radio character to me when I was growing up. To me, he was like a neighbor and a friend...and to this day, that's why in my mind the Lone Ranger rides again!

Farm Families Relish Good Friends

REGARDLESS OF HOW they actually took place, social activities were always important to members of the Lessiter family.

Whether visiting with friends or relatives over tea and cookies, going to someone's home for dinner, playing cards, attending parties, getting together to discuss local government activities, visiting sick people, stitching together an afghan, going to a movie or getting together for church, school or 4-H Club activities, these special relationships were always an important part of life at Lohill farm.

Looking back at parents and grandparents living on the farm, three simple words best sum up their dedication to the needs of others: *"They really cared."*

Caring People

Friends were very important to the Lessiter family generation after generation. They cherished their friends and went out of their way to let people know how much they really valued their friendship.

Family members were always outgoing and keenly interested in what others had to say.

As an example, I don't think Dad ever met anyone with whom he couldn't start a conversation within just a few minutes time.

Before long, he and a stranger

SUMMER FUN. Big parties were very common at the old Lessiter farm house as Grandma Norah and Grandpa Frank loved to entertain. These events often featured a big crowd and Sunday afternoon parties were often held out on the lawn in hot weather.

WHO'S PHOTOGRAPHING? As the photographer snaps this photo of these happy couples, one of the women takes a shot of the photographer. Grandma Norah and Grandpa Frank loved to take trips with other couples to visit special places or friends.

would be talking about some place they'd both been, some piece of farm machinery they had both once owned or some person that both of them knew.

Somehow Dad had that special talent of being able to find the common denominator in just about any conversation.

Through the years, members of the Lessiter family have been blessed with great relationships with many friends. Mom never met anyone she didn't like and could always find something

FRIENDS MEANT EVERYTHING. Mom and Dad also enjoyed many special friendships, such as barbecuing with the Byron Chapin family or sitting with friends at a dinner meeting.

FAMILY WERE FRIENDS TOO. As a youngster, Olive Burt spent many happy days visiting the farm from her home in Oxford. Later, she married Bill McTavish and continued to live for many years in the old family home located in downtown Oxford.

positive to say about everyone.

Her "Golden Friendships"

Many times, she referred to the relationships she enjoyed with lifetime friends in the Lake Orion area, from college, from attending meetings and life in general as her "golden friendships."

"I have known these friends for many years and we have shared our lives together through all of the good and bad moments which come to us all," she once said. "Our understanding of one

ONE THING ABOUT DAD. If he was going anyplace at night or meeting friends, the chances were almost 100 percent that he'd put on a tie and probably a sport coat or suit. That's the way he has been his entire life, as shown here with Mom on their 40th wedding anniversary. Maybe it's because he went out each day to the barn and didn't wear a coat and tie like many other men did when they went to work. He could never fully explain it—he just put on a tie.

WORLD WAR II VETERANS. Carol Kirkpatrick, shown here with Frank on the shore of Lake Orion, was a good family friend who served in the European military efforts during the big war. After the war, he and his wife, Vena, spent numerous evenings playing cards, Scrabble and other games with family members. Next to Mom is Dietrich Ristow, a German exchange student who spent 1957-1958 living with our family.

another is so great that we sense each others' moods and needs. We are secure in the knowledge that there is someone who really cares and is always there whatever comes."

Whether the birth of children, graduations, weddings, family get-togethers, business ventures, successes, achievements and the birth and antics of grandchildren, friends have always been there with Lessiter family members to share the enjoyable news with family members.

They were also there for each other's disappointments, sorrows, illnesses, setbacks and finally the deaths of those we love so much.

"Sometimes we have talked our problems out together," Mom once wrote. "At other times, we have understood the

A Special Friend

Mary Parker had been a good friend of both Mom and Aunt Margaret. In fact, she was with Margaret when she died in New York City at the 1939 World's Fair.

Mary served in the Woman's Army Corps during World War II and was stationed in England, the ancestorial home of the Parker family.

It was a tradition at our house for Mary to drive out to the farm each Saturday afternoon so she and and Mom could have a delightful, yet lengthy, conversation over tea and cookies.

This was always a big treat for the children because either Mom or Mary would always bake cookies that were

shared with all of us who happened to be around the farm that day.

The Godmother of Frank Lessiter, all of the family's children always had a special place in her heart.

Good Friends At Lakefield Farms

Over the years, Lessiter family members were good friends with a number of the people working at Lakefield Farms. This farm surrounded the Lessiter farm on three sides.

Shown above are O.F. and Florence Foster. O.F. served for many years as the general manager of this highly successful diversified cropping and livestock operation.

At upper right is the Bill Hess family. Bill looked after the farm's Oxford and Hampshire sheep. Mike, Stella, Mary Ann and Bill were family members.

In the center is the Kent Mattson family. Kent was in charge of the highly successful Holstein herd at Lakefield Farms. Shown here are Kent, Heidi, Katie, Laurie and Sid. Another child, Karen, was born later.

At right, Kent Mattson and Frank Lessiter exercise Carnation Madcap Butterboy at Carnation Farms in Carnation, Washington. After the Lakefield dispersal, Kent moved west to serve as the Carnation herdsman.

Frank spent a summer with the Mattson family during his college days looking after the bulls in this highly successful West Coast Holstein herd.

DRESSING UP ON THE FARM. Sam Chapin and Janet Lessiter.

Farm Is Always A Good Place For Great Fun!

Through many years and many generations, kids in the Lessiter family had cousins and friends over for lots of exciting days on the farm.

FARM DREAMS. Mike Lessiter.

SMILING FACES. Cousins Tim Tarpening and Janet Lessiter.

COLD WEATHER FUN. Shirley Robertson, Sam Chapin, Janet Lessiter and Frank Lessiter.

meaning of the unsaid word that brings love and understanding beyond measure with just the quiet presence of a friend.

"We know each other's shortcomings and respect them because we see beyond the surface and we understand them. We find each other no less perfect because of these imperfections."

Probably nobody among the many Lessiter family members had more friends than Mom.

"My life is so much richer and more productive because of the beauty and meaning of many friendships which seem to bring out the best in me," wrote Mom. "I only hope that in some measure I have contributed to their lives as well."

Friends Are Important

Through the years, Mom urged us and other farm folks to give our children and grandchildren the desire to find, make and keep friends.

"Teach them to value such a rare gift and to realize it is the giving and receiving that keeps its glow shining like a torch for those which it touches," she said.

"God in his great wisdom saw the needs of people and gave them the blessing of friendship."

Texas-Style Farm Vacations

YOU MIGHT WONDER what the famous King Ranch in south Texas and southeastern Michigan's Lohill Farm have in common.

First, both operations were founded in 1853.

Second, I've listened to Dad recall vivid details more than 60 years later about a 1936 vacation trip that included viewing the King ranch.

After I toured the King Ranch in the mid-1990s, Dad reminded me of his trip to south Texas made six decades earlier.

The main thing he remembers about these cattle barons is the tall fence around the land and the reputation ranch cowboys had for shooting trespassers. Luckily, he stayed on the safe side of the fence!

But there was certainly a big acreage difference between the two family operations. The maximum size of the Lessiter farm was 432 acres back in the 1920s, while the vast Kingsville, Texas, operation founded by Captain Richard King once had cattle and horses running on more than 1 million acres.

Even today, the ranch runs many thousands of Santa Gertrudis (a beef breed developed at the ranch) cows on 825,000 acres. During my visit, we saw one pasture which stretched over 40,000 acres!

Today, they also have a 60,000-acre farming operation

"The Texas vacation resulted in the most expensive grapefruit ever produced..."

---

that includes cotton, grain sorghum and other crops.

The Texas Saga

Like most pioneer farm families, the Lessiters didn't take many vacations. But when they did, they did things up right. This was particularly true with two Texas-style journeys.

Dad recalls the 1936 trip he made with his cousin, Bruce Lessiter, to check out the south Texas property which both sets of their parents had bought nine years earlier. Their auto trip of several weeks retraced a train excursion made by my grandparents, Bruce's parents and two other couples from Lake Orion in 1927.

"They went on one of the trips put together by Texas land promoters," says Dad. "These promoters paid their train fare and showed them land down there that could make them rich growing grapefruit.

"It was the typical Texas land swindle of those days. The promoters would convince Yankees to buy the undeveloped land, charge them to plant grapefruit trees and then bill them a hefty price each month to pay a caretaker to look after everything."

Besides the two Lessiter

brothers, Charlie Howarth and Herb Murphy and the four wives made the trip.

"Everyone but the Murphys bought 10 acres of land and looked forward to getting rich from the sale of grapefruit," recalls Dad. "In hindsight, the Murphys were probably the smart ones."

Over a number of years, each of the three families poured considerable amounts of money into the unprofitable grapefruit project. Even in 1930, Yankee land owners were putting $40 per month into the project for caretaker fees and they never got much grapefruit production or made any money from the land.

"Bruce and I saw how bad it really was during our 1936 trip down there," says Dad.

"I knew nothing about raising grapefruit, but I could see those trees would never make it. I figured it was a lost cause. But then the promoters would change orchard caretakers and tell your grandmother how things were going to get better.

"We urged our parents to get out of the deal. The other two couples either sold out or quit making payments when the monthly costs got too high. They'd even charge extra for irrigation water and other items.

"Sure, there were grapefruit trees on the land, but I think they were shaking down the owners more than the trees. They had also drilled for oil on the land and hit nothing but dry holes."

As a kid growing up in the late 1940s, I remember when a box of free grapefruit addressed to my grandmother would arrive at Christmas time. She always looked forward to these grapefruit grown in her own orchard.

Then Dad would mutter under his breath, "These are probably the most expensive grapefruits this family has ever purchased."

I'll Sell On My Terms!

After Grandpa Frank died in 1949 and the estate was being settled, Dad wanted Grandma Norah to sell the land. He kept after her for several years before she finally agreed to sell the land.

Much to his horror, she listed the land with a broker, but insisted on holding on to the mineral rights.

"I was disgusted and told her nobody would ever buy that worthless Texas land without the mineral rights," he recalls.

"But I was wrong. Somebody finally bought the orchards and she kept the mineral rights."

Check's In The Mail

In 1953, Grandma Norah received a $150 check out of the blue one day. It was a payment for natural gas production based on the mineral rights which she had held on to for the 10 acres of orchard land.

Unknown to Dad, the Pantano Petroleum Company had leased the gas drilling rights for these 10 acres along with many other acres in south Texas. Someplace in this tract of leased ground, they had struck a highly productive gas vein.

In later years, the gas rights were again sold to other companies and the drilling acreage kept getting bigger and bigger.

That was more than 40 years ago and the family still gets a check for this gas. Payments run as high as $150 per year, but it's unexpected income.

So maybe Grandma Norah got the last laugh on what started out as a disastrous purchase from the Texas land promoters back in the late 1920s.

The return from the money poured into those 10 acres of grapefruit has been pretty good.

Yet today's family members don't consider themselves grapefruit growers. Instead, they like to refer to themselves as Texas oil men.

All thanks to Grandma Norah who simply wouldn't listen to Dad's advice!

GOOD FRIENDS, GOOD TRIPS. Grandma Frank and Grandma Norah, shown at left, enjoyed time spent with friends such as at this Gulfport, Florida, picnic.

It Doesn't Get Any Better Than This

By Donalda Lessiter

ONE OF THE MOST enjoyable bonuses of our middle age and retirement years is the opportunity to become loving grandparents. To say it is a delightful experience is a veritable understatement of fact.

Exhausting at times perhaps, but never boring. Those people who said some of the real pleasure of life comes from the time and leisure we have as we grow

SPECIAL DAY. Even though she lived 300 miles away, Grandma couldn't say no when Mike invited her to Grandparent's Day at school. She surprised them and is shown here with Mike, Debbie and Susie looking over a booklet put together by the kids.

THE WISCONSIN BUNCH. Above, Grandpa Lessiter keeps a firm hold on Kelly. At right, Debbie, Mike and Susie pose for a photo, a quick and easy way to let the grandparents see how their grandchildren were growing up hundreds of miles away.

older are certainly right.

But the fact that the real responsibility, the anxieties and concerns for these little lives is primarily those of their parents can also be a relaxing and com-

"I thrive on the adoration our precious seven grandchildren bestow on us..."

forting thought for us old-timers.

I am honest enough to admit I thrive on the adoration and attention our precious seven grandchildren choose to bestow on me

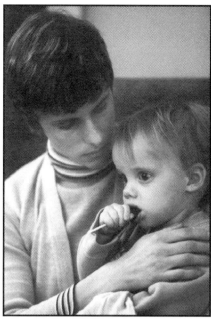

A BUMP. Pam consoles Kelly over a bump on the arm received while playing with her brother, sisters and cousins at the farm.

At left, Debbie and Susie show off new outfits on the first day of school for their grandparents.

and my husband. Before I became a grandmother, I vowed this was an exercise in which I would not indulge. But somehow or other, I gave in to the temptation.

I am guilty of spoiling them

The Treats Of Grandfathering...

A few days ago, our daughter and her husband dropped off our three grandchildren with us while they attended a weekend conference down south.

With all our kids grown and out of the house, we always look forward to these times spent with Alex, Molly and Ryan.

One of Alex's treats is to go to the office with his Grandpa on Saturday mornings. Sometimes, we'll even stop by the small diner in town and order bacon and scrambled eggs.

Last weekend at the office, I watched him race to my model tractor collection, not even bothering to unzip his bulky snowsuit. He sat on the floor and played with those old tractors, disking the brown carpet for several hours, stopping every now and then only to ask, "What are you doing, Grandpa?"

As I chuckled at this "Little Farmer" and the simpleness of this morning I spent with him, I remembered something my Mother wrote nearly 20 years ago. In fact, I went to my old farm trunk that came from England back in the mid 1850s to look for it and I hope you enjoy it.

***"Grandchildren are God's way
of rewarding us for growing old."***

THE BIG HILL. Mike takes a run down the hill by the big barn as Trista and Grandpa Lessiter watch. At right, Mike checks the pump as Grandpa Lessiter fills the pickup truck with gas.

when it comes to the cookie jar, a well-stocked candy shelf, story-telling, cards, letters, unexpected gifts and the ever-open listening ear.

Some Strange 'Friends'

Our kitchen bulletin board is

"I love to share the many unpredictable capers of our seven grandchildren.."

constantly updated with new snapshots, drawings, letters and even poetry. What priceless and loving gifts they are—and how gladly received. And as my

HI COUSIN! Kelly, Andy and Mike enjoy getting together during summer vacation.

278

LATE AFTERNOON DELIGHT. Grandpa hangs on to Andy as Grandma looks on from the screened in porch which was an important part of one of the Lohill Farm farm homes.

PLANTING CORN. Mike demonstrates how his ancestors used to seed corn with this antique seeder.

friends know, I love to tell some of the more unpredictable capers and profound statements of our grandchildren.

Conversations with these little, highly imaginative minds might verge on the ridiculous or idiotic to someone else. But to those of us who participate as grandparents, they are very real indeed.

We visit about monsters, the pink cow with the green shoes, fire engines, garbage collectors, storybook characters, tea parties and even Grandpa's lack of hair.

TIME FOR FUN. Mike enjoys a walk down the farm lane while Kelly rides one of the farm bikes.

MOST BEAUTIFUL BOUQUETS. That's how Grandma Lessiter described bouquets of flowers and weeds picked by her two oldest granddaughters, Susie and Debbie Lessiter.

DANDELION BEAUTY. While it was tough for Dad to see the beauty in yield-robbing dandelion-covered corn or hay fields, he made an exception with granddaughter Katie Roberts.

enjoy. Sometimes, I am even invited to share a grubby cookie or a once-licked gumdrop.

One of my favorite pastimes over the years has been creating articles involving our grandchildren in simple everyday stories. My efforts along these lines may not be great by literary standards. But to my grandchildren, I've learned to accept imaginary gifts with enthusiasm and surprise. My bouquets of dandelions or weeds are beautiful and deserve being displayed for all to

HAPPY FARMER. Andy Roberts enjoys a good meal served by mother Janet. Unlike his cousins, he was able to enjoy rural life to the fullest by growing up on the farm.

GRANDMA'S LITTLE GIRL. Trista was a daily visitor at the Lessiter house. When Mom and Dad left to teach each morning, she practiced the piano while waiting for the school bus.

they are Pulitzer Prize material…and that is sufficient for me.

I love the inquisitive minds and the active imagination of each child. Reading is a patient study in "Who's Who?" The in-

spired sidetracking of thoughts is of amazing originality.

I am a captive and responsive audience to the tales of the amazing powers of Murphy, the Irish Setter-Black Lab mix who is the constant companion of Mike. As Mike tells it, Murphy is the equal of Babe, the Blue Ox of Paul Bunyan fame. Likewise, the children's imaginary friends are special friends of mine as well.

200 Years Old?

When our two oldest granddaughters, Susie and Debbie, were 3 and 4 years old, I invited a blow to my ego when I replied to their question of how old I was by asking, "Well, how old do you think I am?"

Their Grandfather is still chuckling at little Susie's candid answer of "200!"

WHO'S BLUFFING? After a hard day of play on the farm, Mike Lessiter plays cards with Grandpa Lessiter.

Later that year, we found out by chance that we were real "celebrities." There was that inevitable children's game of "Mine Is Better Than Yours," which we overheard from our son's patio.

The Teeth Trick

We were surprised to hear Debbie, backed by her younger sister, Susie, out-top all the neighborhood kids by contesting, "That's nothing. My Grandpa can take his teeth out and Grandma has silver in her mouth. If you don't believe it, they'll show you."

And show them we did—as we staunchly defended the integrity and newfound prestige of our grandchildren. To an attentive audience of neighborhood

"She invades the cookie jar and candy shelf whenever the mood suits her..."

children, Grandpa dutifully clicked out his partial plate, not once, but several times, and I willingly displayed my numerous silver fillings.

My magic was not overwhelming, but Grandpa's tooth magic remained a mysterious talent for many years while the girls were growing up. And as Trista, Kelly, Andy and Katie appeared on the scene, they were

LOVE THE FARM. Alex Hansen is a great-grandchild who enjoyed a visit to Lohill Farm.

GRANDCHILDREN ARE GREAT! Talking about his grandchildren and great-grandchildren always brings a smile to the face of Grandpa Lessiter.

properly introduced to Grandpa's special talent. The fact that they unsuccessfully tried to duplicate the feat only added to his reputation.

Daily Visitors

Trista, the young child of our daughter who lives next door to us on the farm, is a daily visitor to whom nothing in our house is sacred.

She invades the cookie jar and the candy shelf whenever the mood suits her.

She investigates the mysteries of our desk and magazine rack.

She bestows her talents on our piano.

She reminds me frequently that the "no-no's" on the five-story antique "what-not" shelf hold much promise for her when she is a "big girl."

Conveniently, however, she "can't remember" at times. At other times, her memory is nothing short of remarkable.

Having grandchildren on the farm represents many precious moments. Trista plays the piano at our house and having the bus

---

"I strongly recommend grandparenting and I know you will love it..."

---

stop across the street, we are blessed with special time with the kids.

Little Andy takes advantage of this opportunity, eating two breakfasts before school. After awakening and eating his favorite cereal at his house, he walks down and shares one of his Grandpa's famous soft-boiled eggs until the bus arrives.

There's a saying, "Grandchildren are God's way of rewarding us for growing old." That's the only way to describe it—rewarding.

Life's Biggest Rewards

Grandparenting is truly a job—a two-way one. As our daughter once said about the role: "Where else can you go where you are always so important, so loved and so wanted?"

She is right. It is a real shot in the arm to us grandparents to feel wanted, loved and important. Taken with a liberal dose of "no interfering with parents" and an ever-so-slight bit of "making them mind," grandchildren are a pleasing and palatable addition to our lives.

The hugs and occasional "I love you's" have indeed made Grandma and Grandpa willing slaves to our grandchildren—the kings and queens of our hearts.

I strongly recommend grandparenting to all older couples.

Try it, and you will find that you love it.

Special Letters To Grandma Lessiter

Mom Always Enjoyed Her Family...

Most of all, Mom really loved and enjoyed the time spent with family and friends, especially those precious moments spent with her seven grandchildren.

While four of her grandchildren grew up 300-miles away, it was always a special time when they came back to the family farm.

That's why I want to always remember Mom with the two letters shown here which our son wrote to her when he was only 8- and 10-years old.

These letters point out something Mom always thought some farm folks tended to take for granted...the many good things that go with country living regardless of the season of the year.

STAYING IN DAD'S OLD ROOM was always a highlight for Mike on the exciting trips which the grandkids made back to the farm.

Dear Grandma:

IT WILL ONLY BE a few weeks before our family will be coming to spend a few days with you and Grandpa on the farm in August. So I hope you are getting my Dad's old bed ready for me.

As you know Grandma, I love to sleep in the bed that was Dad's when he was a boy like me. I like to lie there in that old bed and think about how lucky he was to grow up on a farm. That sure had to be fun for him!

Lying there, I think about how

"FOR SOME REASON, I seem to sleep better in Dad's old bed at the farm than I do in my own bed back home in Wisconsin."

Dad used to look out the bedroom window and see the white barns and the milk cows grazing in the nearby pasture. And how he would get up at 6 a.m. to help Grandpa feed and milk the cows before heading to school. He sure doesn't get up that early anymore!

I like to lie there in bed and wonder which of the books still

"AT HOME, I NEVER want to get up in the morning. Here on the farm, it's just not a problem."

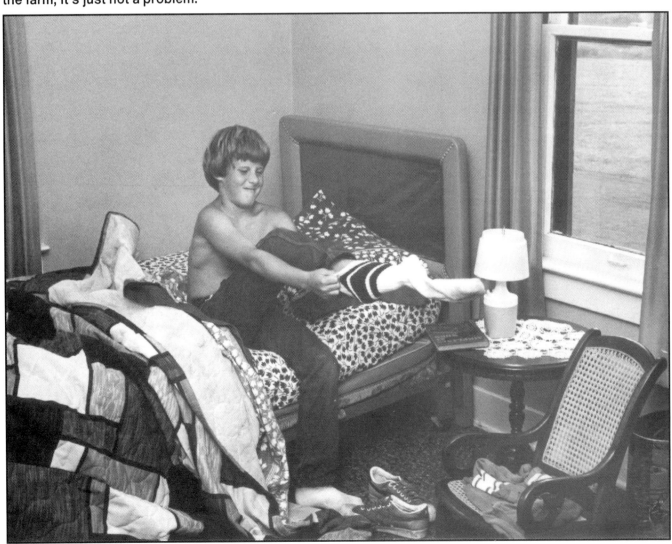

on the shelves my Dad read when he was growing up.

I like to wonder what my Dad used to think about after he went to bed at night. Since he raised lambs, did he count sheep when he couldn't sleep?

But most of all, Grandma, I like to lie there in bed and pretend I'm going to grow up and be just like my Dad. And I want you and Grandpa to think that I'm just as good as my Dad, too.

So Grandma, please make sure I get to sleep in my Dad's old bed again this summer. I'll be seeing you in a couple of weeks.

"EVEN WHEN IT'S EARLY, I'm always ready to get up and help Grandpa with the farm chores."

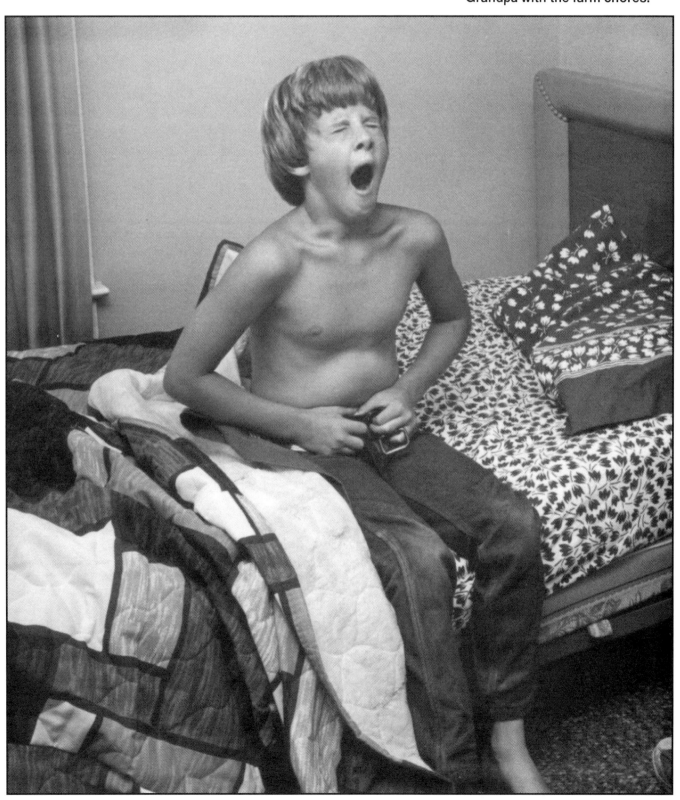

"JUMPING DOWN is always the neatest way to get down off the tractor after I've driven it."

Dear Grandma:

WHEN I TOLD MOM I was going to write and thank you for my extra special week at the farm, she asked me why I like to visit you so much.

I told her there were a whole bunch of reasons:

1 It's a neat place to live because there is so much to do. If I lived on the farm, I'd work hard every day just like Grandpa does. I like to work with my Grandpa.

2 Farm chores are a lot more fun than city chores like taking out the garbage, cleaning the table and picking up after the dog. I like feeding the cattle and pitching manure lots better.

3 Driving your big tractor is great. I still have the note on my bulletin board that Uncle Neil wrote a note to my Dad when I was 6-years-old that said, *"This is proof that your son, Mike, drove the tractor all by himself."* Of course, Uncle Neil was there on the seat helping me a little.

4 Having hay fights with my three sisters and cousins is a riot. But Mom didn't smile when she had to get all the hay seeds out of my hair. I love climbing up high on the hay bales and then jumping down.

5 It's "neat-o" to try and dump a big bag of feed into the chicken feeders. But I guess I'm not big enough yet, since most of the feed ends up on the ground.

6 Another reason I like to visit the farm is to eat your chocolate chip cookies, Grandma. They're yummy. They are the best cookies anyone makes. But please don't tell that to my Mom because hers are pretty good, too.

7 I like riding the pony Strutter. He trots too fast for me, but it's really fun to try and hang on as he runs down the lane toward the lake.

Grandma, do you remember the time Strutter tossed my sister

> *"Do you remember the time when Strutter tossed my sister off in the snowbank?..."*

Susie off in the snowbank when we visited you 2 years ago at Christmas? Susie sure got mad when I laughed.

8 It's fun to squeal and grunt like the pigs do. I still remember how loud a baby pig squealed when I picked him up. Then the sow got up and started grunting real loud. That's when I put the pig down.

9 I like climbing ladders in the barns and silos. Even my

"IT'S ALWAYS FUN TO DRIVE the tractor, even if Uncle Neil sits there on the seat with me."

Dad was surprised last summer when I climbed to the top of the old 40-foot silo.

He said it was okay, but there was no reason to tell Mom or Grandma about it. Oh, oh—now you know!

10 I think pitching manure is fun. Dad says that's

"IT ISN'T REALLY THAT HARD to drive the tractor, but I'm still glad Uncle Neil is there to help me."

"AUNT JANET AND I try to catch a rabbit that is under the corn crib. I hope it won't bite because that can hurt. Grandma, Aunt Janet and I finally caught the rabbit. You're so lucky to be raising farm animals."

"GETTING A RIDE on the trailer with my sisters and cousins is always lots of fun."

"FINDING A WAY into one of the old barns is always fun, and so is seeing what we might find inside."

'cause I just haven't done enough of it. I guess the real fun is when Uncle Neil lets me drive the tractor after I help him load the spreader.

I sure don't think so, but Dad always says I got the worst of the deal.

11 We've got fancy swings at home, but they sure aren't as much fun as swinging on the old tire on your maple tree out by the barn. I can sure swing high, yell all I want and make all the noise I want. I can't do that back home.

12 Dad's old bed is always my favorite place to sleep

> *"I think about all the neat stories you and Grandpa have told me about when my Dad was growing up..."*

at your place. I like to lie there and pretend I'm lucky enough to live with you on the family farm.

I think about all the stories you and Grandpa have told me about when Dad was growing up on the farm. You know, Grandma, I sleep better in his old bed than any other place. Even better than I sleep in my own bed back home.

13 Best of all, I like to come to the farm to see you and Grandpa. It sure is a great place to have fun. My Dad says maybe someday we can all move back to your farm. I hope so and before too long.

Your farm sure has a lot of great things for me to do that my sisters and I simply can't do here in the city.

"GRANDMA, GETTING TO SPEND A FEW DAYS each year at the farm with you, Grandpa, my aunt, uncle and cousins is always one of the real highlights of my entire summer. I wish we could live back on the farm. Even with all the hard work that has to be to be done every day, it's still a really neat place to live."

Township Service Was Long-Time Family Tradition

ORION TOWNSHIP now has a population of around 50,000 people in a highly suburban area located 35 miles north of Detroit, Michigan. The township budget runs into millions of dollars.

But when the first township supervisor, Jesse Decker, was appointed in 1835, his salary was only $2 per year. The assessed valuation for the township was only $21,530. Township expenses that first year ran $36.96 and tax receipts were $38.68.

Family Tradition Starts

The first member of the Lessiter family to serve as township supervisor took office in 1918. And Dad wrapped up the Lessiter family legacy of serving as township supervisor when he retired in 1972.

The first supervisor in our family was Dad's Uncle Floyd. He was later appointed Oakland County Poor Commissioner to handle welfare cases in those days.

When he resigned as township supervisor to take the Poor Commissioner's post, they appointed Grandpa Frank, who had served 2 years as township treasurer, to take his place.

Grandpa Frank served nearly 10 years as township supervisor

"That first year, the total assessed value of all township property was only $21,530..."

until 1927. He was also an officer of the local school board, served as Justice of the Peace for 18 years and was an active Democrat in local politics.

Back in those days, the supervisor's part-time job was done from a desk in his home. All township records were moved to the supervisor's home when he got the job and there was always plenty of work. Dad says Grandmother Norah did much of the bookwork.

Figure, Figure, Figure

"I remember when Dad went up to the railroad station in Oxford and was talking to Mr. Oliver about all the tough work in figuring the township taxes," recalls Dad.

"He let your Grandfather borrow an adding machine for a couple of months, which really helped in figuring the township and county taxes that year. Up until then, your Grandmother and Grandfather were calculating all the taxes by hand."

From 1927 until 1946, nobody in the family ran for elective office, although Grandpa Frank and Grandma Norah and later Dad were all involved in local political activities. In fact, Grandpa Frank insisted that attending the township meeting was a "must" every year.

Dad served on a few township committees and put in a long stint as President of the Lake Orion School Board. But for a number of years, nobody in the family had sought political office in township politics.

Dad first ran for township supervisor in 1961 against Ferris Clark who had held the office for a number of years. One of the election campaign criticisms of Dad was that he was a full-time farmer who was trying to take away the job the other candidate had held for so long. The major question on the minds of voters was what Clark would do if he lost his job.

I still remember something that happened during the middle of the spring election campaign while I was home from college on Easter vacation. A grass fire got started in one of our fields that spring and Dad and I really worked hard to get it under control—to no avail.

He really didn't want to call the fire department for help since he figured someone would think he was seeking something extra from township officials. Or that it would draw attention to his full-time farming career.

But when we couldn't control it, he finally called the fire department who swept into the field with a couple of trucks and had no problem getting it out.

And I'm sure it didn't cost him any votes. In fact, he was the first Democrat to be elected in many years to this township post.

Time For A Change

Voter attitudes had changed as citizens saw major concerns in township development that weren't being properly handled or even discussed.

Dad served from 1961 until 1972 when he was 64 years old. That's when Mom told him it was time for a change and he didn't run for re-election.

By 1961, being township supervisor was definitely a full-time job. The township was really growing, plenty of new homes were being built and significant decisions had to be made regarding the way the township was to be run and the need for future planning.

Fair, But Firm

As you would guess happens in politics, Dad had a few run-ins with folks who wanted things done differently. One of his main attributes was that he would always listen to everyone's concerns and had a well-earned reputation for being fair with everyone. Orion citizens didn't always agree with the eventual decisions, but they always had the opportunity to be heard and to voice their ideas and opinions.

No Pre-Fab Homes?

One argument with a few voters took place when the Keatington complex was being developed in the mid 1960s on the old Scripps family's Wildwood Farms. There were some very nice houses going up in this complex and a few residents were upset when they learned pre-fabricated homes were going to be built on a few lots. They felt these "pre-fabs" would not blend well in the subdivision and would lower the value of their homes.

Dad had seen these houses in other locations and knew this wasn't the case. So he told residents he was going to allow two of the pre-fab homes to be erected so everyone would see what they looked like. The residents weren't happy, but went along with the idea.

As soon as the two homes went up, the commotion was over. The owners realized these houses were probably even better than the ones they had built and would likely increase the value of other homes—rather than lower the value.

Dad wanted to buy one of the lake lots in Keatington after the official October 27, 1964, groundbreaking, but Mom wanted to keep living on the home farm and she also didn't think it was a good idea for a politician to buy a lot from a developer he was dealing with on a regular basis. So they didn't buy the lot—which had a price tag on it in those days of $14,000 and could probably be sold today for over $80,000!

The developer later honored Dad for all the work they had accomplished together in this big development by naming a street after him. Over the years, we've taken our children and

LESSITER DRIVE. This Keatington street was named after Dad when he was Orion Township Supervisor. It was the developer's way of honoring him for the work involved in getting this project underway.

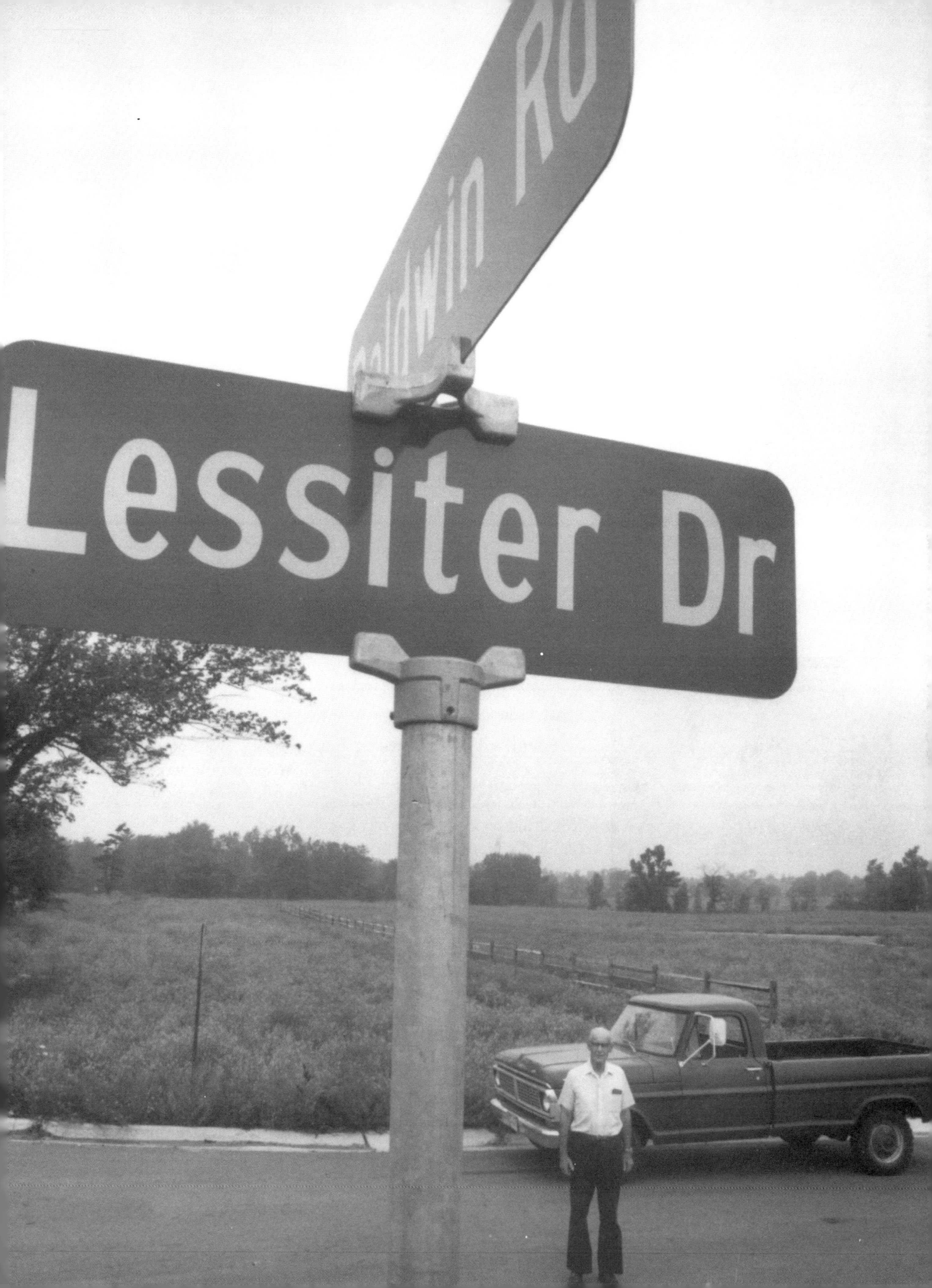
Lessiter Dr

grandchildren to see the sign on "Lessiter Drive."

Famous Resident Run-In

In those days, Dad also did all the tax assessing work, a job which now has a separate staff because of the the township's phenomenal growth. As township supervisor, he also served as a member of the township Board of Review which met once a year to hear complaints from citizens who thought their taxes had been increased too much.

The previous year, Dad had sharply increased the assessed value of Teamster's President Jimmy Hoffa's home, located on Square Lake. Hoffa had made many improvements and additions to his home in recent years.

Plus, as a means of showing their gratitude for the job he had done for the Teamster's organization, union officials arranged to have his driveway blacktopped at no charge that summer. Hoffa's driveway was three-quarters of a mile long, so the value of this blacktopping was extensive!

When Dad went to the Hoffa residence to make a tax reassessment, it was quite a different property than had been assessed years earlier. As a result, he

---

"Jimmy Hoffa was mad about Dad's increased assessment of $50,000 on the property..."

---

jumped the assessment from what he remembers was $85,000 to over $135,000.

When the Board of Tax Appeals met that spring, an unhappy Jimmy Hoffa was among the dozen people pleading for a tax cut. But he didn't get a single cent of reduced taxes!

Handling "Job" Stress

Being township supervisor was a stressful job, yet Dad loved the challenge of working on local government affairs.

When he became township supervisor, he still had about 3,000 laying hens and it was tough handling both jobs. But he didn't want to give up farming until he was sure he really liked the job and had the confidence that he would be re-elected.

Deciding he enjoyed the work, he sold the poultry flock. For the first time in 110 years, the family didn't have any livestock on the farm!

Dad eventually rented out some land, cropped other fields and sold hay and straw—1, 2, 5 or 100 bales at a time. While he didn't raise small grain crops for straw, Dad bought baled straw from farmers located further north and sold it to customers.

It was hard for a farmer to believe, but there were times when straw was priced higher than hay because of the short supply and heavy demand. Much of the baled straw went for archery targets, pet cages, gardens, horse stalls and for curing concrete in cold weather.

Dad continued a daily ritual which helped him relieve the stress of the township government job and keep fit as a fiddle.

He always got up early and went to the barns for an hour's work each morning before going to the office. As is likely the case at your place, there was always something that needed to be done on the farm.

Dad and Mom joked about how this early morning ritual allowed him to better handle the stress of the office job and avoid a heart attack. Plus, it helped him keep his sanity to carry on an early morning work pattern that had been a tradition all of his life. Besides, he could never sit still!

50 Years Of Records

In his career as township supervisor, Dad remembers that one of the more delightful aspects of the job was going through old important township records and finding a land plat or other key document signed by Grandpa Frank when he was supervisor back in the 1920s.

As a kid in the old farm house, maybe Dad was even looking over Grandpa Frank's shoulder as he signed a document...one he was now holding in his hand 50 years later.

54 YEARS OF CHANGE. Several members of the Lessiter family served as Orion Township Supervisor during periods from 1918 until 1972. During this time span, family members saw many changes, including the tremendous amounts of traffic on this part of Baldwin Road running past Lohill Farm.

Family's Second Centennial Farm

FEW FARM FAMILIES have ever owned even one Centennial Farm. Yet the Lessiter family owned two such farms for 105 years.

Besides the home farm which was started in 1853, Grandma Norah Wiser's family had moved on to a farm at Seymour Lake in 1831. This farm was located 3 miles north and 2 miles west of the home farm.

The Seymour Lake farm was originally purchased in 1831, so it was actually in the family for 22 years before the home Lessiter farming operation got underway. Grandma Norah was born and raised on the farm.

Rough And Rocky

The farm consisted of 80 acres—a very rocky, very hilly 80 acres. The soil was so rocky and so lacking in essential plant nutrients that it never produced very much in the way of profitable crops.

It always amazed later generations how Grandpa Wiser actually made a living off the farm. But there was definitely one thing in his favor—the family was quite frugal in the way they lived.

Headed To Town

"I remember hearing the story many times of how Grandma Wiser and my mother went into

THE OLDEST FARM. The Seymour Lake farm was purchased in 1831, 22 years before the Lessiter family farming operation was started.

297

A STEEP, STEEP HILL. Even in the summer, it's always a tough, breath-taking climb up this hill to the Wiser family burial plot in the Oxford cemetery.

But what I remember most is the time in December of 1945 when Grandma Wiser died just after Christmas at the age of 100. It was snowy and very icy that day, which made this one of the toughest walks anyone ever made up this hill for the graveside ceremonies. With all the ice on the hilly paths, it certainly was a rough trip for the pallbearers.

Oxford one day and saw a house for sale," recalls Dad. "They bought the house without having my Dad along since he was showing cattle at the fair. Boy, was he ever surprised when he got home."

Grandma Norah's folks moved off the farm via a 1-hour cross country wagon trip in 1909 when they retired to a house still standing on Broadway Street in Oxford. Later, they moved to a house on Washington Street in Oxford.

The Seymour Lake farm was leased out for years until it was sold in 1958. Today, it has been subdivided and a number of houses stand on the old farm.

Grandpa Wiser died in 1912 and Grandma Wiser outlived him by 33 years, celebrating her 100th birthday 2 1/2 months before her death at Christmas time in 1945.

THE SEYMOUR LAKE CHURCH. Still used today, this is the old Methodist church where Grandma Norah and Grandpa Frank were married in 1895.

THE FAMILY PLOT. Located high on a hill in the old section of the Oxford cemetery, the view from these cemetery lots is superb, especially in the fall when the leaves are in full color.

THE OXFORD HOUSE. Grandpa and Grandma Wiser moved off the Seymour Lake farm to this house in Oxford in 1909. He died in 1912, but Grandma Wiser reached 100 years of age in October of 1945 before she died 2 1/2 months later.

501...

HOW MANY families with one address do you know with two mailboxes for two towns?

Dad and Mom had two mailboxes for our 501 Baldwin Road address for more than four decades. The Clarkston mailbox was the first one in front of our house. But since we seldom went to Clarkston, Dad preferred a Lake Orion address and had a second mailbox located one-half mile north of the farm.

"You kids went to the Lake Orion schools, our family did all of our business in Lake Orion, your Grandpa was on the board of directors for the Orion State Bank and most of our family friends were located there," recalls Dad.

"Everyone figured we had a Lake Orion address and would send us mail there—even though we had a Clarkston address.

"Before we added the Lake Orion mailbox, people would get mail returned to them stamped 'no such address.' Yet they knew we were still there—they'd just seen us in town."

The solution was to add the second mailbox to get mail from Lake Orion's Post Office.

The Canadian Farm Family

DONALD TAYLOR spent eight years trying to eke out a living as a western Canada dryland wheat farmer and owner of a livery stable.

Unfortunately, neither his daughter, Donalda, his grandchildren or his great grandchildren got to enjoy his company.

Taylor grew up as part of a large farming family near Rodney, Ontario. The family farm was known as the first concession of Aldborough and was located along the Thames river north of Rodney, Ontario.

Donald was the son of Duncan and Catherine Taylor who had been married on February 5, 1857, in St. Thomas, Ontario. He was the tenth child among six girls and six boys born into this

TAYLOR FAMILY BIBLE. Generations of family records are recorded in this Bible used by the family for more than 100 years.

EARLY-DAY LOOK. This was the Donald Taylor family residence on the Aldborough farmstead.

farm family from 1857 to 1879.

His father, Duncan Taylor, was born on December 9, 1819, and lived to be 76 years old. Hailing from Scotland, he spent the first 57 years of his life at Dorchester, Ontario, before moving to the Rodney area farmstead in 1874.

During their adult years, the dozen Taylor children were involved in farming, carpentry, bookkeeping, implement business, retail store work, office management and nursing.

Through the late 1940s, the two youngest sisters, Anne and Belle Taylor, lived together on the old family homestead before moving into Rodney, Ontario, in the late 1940s. Belle had previously worked in Detroit, Michigan, for a family-owned lumber company. She came back to the Ontario farm to take care of her sister, Catherine, who was bedridden for many years.

Go West, Young Man!

Donald moved to a farm near Nanton, Alberta, in 1908.

"The Canadian government in those days would give you a certain amount of money if you went west and homesteaded farm land for a minimum of five years to help develop the western provinces," recalls Dad. "As the family told the story, Donald went west to seek his fortune, but found he didn't like farming out west in the dryland areas of Alberta all that much.

"So he ended up moving off the land and buying a livery stable in Nanton."

Several years later, Janet Mur-

RIVER BANK LAND. The acreage flows back from the farmstead toward the Thames River, named after the famous river in England.

ray of Silverwood, Michigan, moved to the area to keep house for one of her uncles. She met Donald and they were married on July 22, 1911. They had a daughter, Margaret who was born in 1912.

On July 20, 1914, Donald Taylor died of an apparent heart attack. His daughter, Donalda, was born 10 days later and she was named for the father she never had the chance to meet.

MODERN LOOK. The same house still looks very similar in modern times. At right is the house which Anne and Belle Taylor moved to in the nearby town of Rodney when life became too rugged on the farm.

Shortly after Donald's death and Donalda's birth, Janet Taylor and her two daughters moved back to the Detroit, Michigan, area where two brothers worked for the Hudson Motor Company. She soon started a boarding house in Detroit and met Sher-

OLD, OLD BARN. The Taylor family built this barn many years ago. It is still used today by the Alderton family who purchased the farm over four decades ago.

TAYLOR PLOT. Located east of the old farm, the Taylor cemetery plot is the burial ground for several generations of the family. Mike and John Lessiter check the engraving on the tombstone of Donald Taylor who died in Alberta at the age of 40. He is the Taylor link to the Lessiter family.

man Tarpening who worked in Detroit. He was a boarder who actually ended up marrying the landlady.

They were married in 1919. Shortly after, they moved to Lake Orion to a home on South Broadway located just a block from the lake. Their son, Sherman, was born in 1923.

While Grandpa Tarp was an "adopted" father to the two girls, he was a wonderful father. No one in the world ever would have guessed he was the two girls' stepfather as they loved and treated him as their own father.

During the years spent in Lake Orion, he held many jobs—an engineer on the interurban trains that ran between Detroit and Saginaw, a factory worker in a

FAMILY BURIAL INFORMATION. Note the sketch and receipt for a tombsone ordered by the Taylor family in 1902. Death notices dating back to 1895 were found among pages of the family Bible.

tank plant during World War II, a cemetery grave digger, a worker at Jacobson's Greenhouse located across the street from the house and the 7 p.m. to 6 a.m. watchman and heating engineer at Jacobson's for many years, keeping the flowers from freezing in the greenhouses.

Health Concern Arises

Margaret Taylor died suddenly while on a trip to New York City to attend the 1939 World's Fair accompanied by long-time family friend Mary Parker. Only 27 years old, she died of an apparent heart attack.

She had served seven years as secretary to the Lake Orion Superintendent of Schools.

While there is no absolute scientific proof, a number of medical doctors today speculate that the apparent heart attacks of Margaret and her father, Donald, may not have been a heart attack

ON-FARM BURIAL GROUND. Surrounded by growing corn and soybeans, a small fenced-in burial ground contains tombstones for people unrelated to the Taylor family which date back to the mid-1800s. As is the case with many of the early-day burial grounds, this one is definitely showing its age.

FERRY CROSSING. The easiest way to get from Lake Orion to Rodney, Ontario, was to take this ferry on the St. Mary's River from Roberts Landing to Port Lamberton. It was much easier than going through the tunnel or across the bridge in Detroit or crossing the Port Huron bridge.

at all.

Based on evidence that showed up in later medical files for three generations of the family, starting with grandma Tarpening, it is believed that a unusual chromosome defect may have been caused by what is now known in medical terms as pheochromocytoma. This problem eventually has shown up through four generations of the family. Fortunately, through surgery and monitoring family members with this condition, there is a much better prognosis today.

TARPENING CONNECTION. Grandpa Tarpening and granddaughter, Janet, enjoy a mid-summer family picnic. Aunt Grace, who kept house for him for a number of years, is sitting on the picnic bench. Below are the three children: Margaret, Sherm, a World War II Navy man who served on the U.S. Cushing destroyer, and Donalda.

Visiting The Old Taylor Farm In Canada

FOR MANY YEARS, my Mother told my sister and me about the numerous trips she had made as a child to the farm owned by her two aunts near Rodney, Ontario. Here are a couple of the memories she remembered best.

Two very homey memories Mom had of visiting her aunts at the Canadian farm, which had been in the Taylor family since 1874, were their hand-made quilts and the old tin button box.

Great Quilts

The numerous quilts were beautiful and often featured unusual designs. But Mom's favorites were the patchwork quilts which were made from small scraps of material from which her aunts had sewn dresses and aprons or those scraps which friends and relatives had contributed for quilting.

Yet it was not the beauty of the quilts that held Mom's interest as a child. Instead, it was the many storys behind the scraps and the incredible memories her aunts possessed while recalling the source of each small piece of material.

One quilt patch might be a piece of material from

"Many were people Mom had never met, but she knew their stories and histories well...:"

someone's bridal gown while another scrap came from material used for a dress for a special party or event. A pink scrap was from someone's baby dress and the white scrap was from a christening outfit donation.

The blues usually came from Aunt Maggie's scraps while the striped ones came from shirts which had belonged to Mom's uncle.

A very special quilt patch was the one that came from a scrap from Mom's own tenth birthday dress.

As her aunts would point out numerous quilting patches, the conversation frequently brought out some interesting, most unusual and sometimes rather revealing tidbits of information about various people.

Many were people who Mom had never seen, but she still knew their stories and histories very well.

The Button Box

Mom used to tell us how the old button box at Aunt Belle's

TAYLOR FARMSTEAD. Located at the "first concession of Aldborough" which is a few miles north of Rodney, Ontario, this acreage became the Taylor family farm back in 1874.

and Aunt Annie's farm home was always an intriguing source of joy. When Mom as a visiting youngster was bored and had nothing else to do, Aunt Annie would often ask, "How would you like to look at the button box?"

Mom told me she never tired of looking at the hundreds of buttons kept in this hand-painted tin box. And she admitted many times to my sister or me how the quilts, with their scraps of material from which clothes had been made, never held quite the same fascination as did the buttons which also had their own stories.

These buttons seemed to make the people who had worn them come alive as Mom held the button in her hand and listened to the story told about it.

There were buttons of every shape, size, color and texture. There were bone ones, glass buttons, leather ones, cloth-covered buttons, wooden ones, silver and gold buttons, brass ones, pearl buttons and many more unusual types and shapes.

There were tiny buttons from baby clothes, beautiful and

> "Each visit seemed to have additional buttons to be admired and sorted..."

dainty buttons from party gowns, a bronze button from Earl's Army uniform, black buttons from family member funeral attire and the button that came from Grandpa's wedding frock coat.

There were overall buttons, shoe buttons, sturdy underwear buttons, heavy-duty coat buttons, tiny glove buttons and fancy buttons for special trimming needs.

Mom used to tell me there was no end to the button collection. Each visit seemed to have additional buttons to be admired and sorted.

Fond Memories

Years later when Mom used to search around to find a replacement for a missing button for my shirt, she would recall the old button box. She would tell me how she always felt a nostalgic loss of that old and unique custom of keeping all the family's buttons in an old tin box.

The Farm Next Door

IT SURROUNDED our farm on three sides and stretched for almost a full mile along Baldwin Road. And the 1,200-acre Lakefield Farms certainly had a big impact on our own family's farming operation.

The formation of this large-scale farming operation east of Clarkston, Michigan, and the owners' desire for valuable land even put a sizable amount of money into the pockets of my grandparents in the late 1920s.

Detroiters Become Farmers

The owners eventually owned all of the fields around Dennis Lake—thus the name Lakefield Farms. It was founded in 1914 by John Lambert and his son-in-law, Oscar Webber.

This was in the era when many large-scale, multi-species, hobby-style livestock farms were developed—often bankrolled by highly successful businessmen wishing to enjoy a touch of agriculture.

One of Detroit's leading businessmen, Webber served for a number of years as the general manager of the gigantic 15-story J.L. Hudson Department Store, which covered a full city block in downtown Detroit.

LIVESTOCK BARN. Built by Dad's Uncle Floyd, this structure replaced an earlier building which had burned. At Lakefield Farms, the barn was used for draft horses and later for hogs. Feed was stored upstairs with entrance from a second story banked area on the other side of the barn.

Under the Lambert and Webber ownership, Lakefield Farms not only became famous for its Holstein dairy cattle, but also for its Oxford, Shropshire and Hampshire sheep and Duroc hogs.

The long-time farm manager, O. F. Foster, came to Lakefield Farms in 1924 as the hog man and became the farm manager a few years later.

"He is considered by agricultural leaders as the most outstanding farm manager in

COMFORTABLE COWS. Kent Mattson helped develop the Lakefield Farms Holsteins into one of the nation's finest herds for both milk production and type.

"They had their minds made up—they wanted to own land all around the lake shore..."

Michigan," wrote noted auctioneer C.B. Smith in the catalog for the Lakefield Holstein dispersal in 1956.

As Dad recalls, Lambert and Webber had their minds set on controlling all of the farm land surrounding Dennis Lake—now known as Heather Lake. And they really didn't care what it cost to accomplish this goal.

The first land purchase was the Cobb farm, located east of the house where Heather Lake Estates now has its general office.

Couldn't Afford Not To

Dad's uncle, Floyd (Jay) Lessiter, farmed the land stretching along both sides of the intersection of Baldwin and Clarkston Roads. When he married Lillian Walter of Clarkston, he moved into the house at the corner of Baldwin and Clarkston roads. The house had been built around 1875.

Dad remembers when the

O.F. FOSTER. Starting out as the hog man, he later served as farm manager for many years at Lakefield Farms. The Foster house was built for his family when they came to the Clarkston, Michigan, farm in 1924.

310

HIGH PRICED LAND. In the mid 1920s, John Lambert and Oscar Webber decided to buy all the land surrounding Dennis Lake.

When the price got high enough, members of the Lessiter family sold some of the best land they had so the two owners could control the lake. There were also several high producing apple orchards at Lakefield Farms.

original barn on the southwest corner of Baldwin and Clarkston Roads burned and how his uncle bought a woodlot down by Webber School, put a sawmill in the woods and cut lumber to build the new barn.

When the Lambert and Webber offer for the land got high enough in the early 1920s, the family sold out and moved a mile west on Clarkston Road to another farm.

"Uncle Jay also owned a cranberry marsh down next to the lake and he was paid top dollar

BEAUTIFUL BARNS. The two big L-shaped dairy barns at Lakefield Farms were once among the finest and most modern to be found anywhere in the country. Suffering from a lack of use and attention for more than 40 years they've fallen on hard times as have similar barns on land no longer being farmed.

for that land, too," says Dad. "As a kid, I didn't like to go there because it seemed like the ground would shake as you

The "Show Place" Farms...

Lakefield Farms was among a number of large-scale hobby-style, tax-deductible farms owned by businessmen in the early to mid parts of the century in Oakland County, Michigan.

Located just north of Detroit, the county proved to be a mecca for these farms. But by the 1950s and 1960s, most of these farms had disappeared due to land inflation, estate planning, rising costs, a changing tax picture and the growing demands of suburbia.

Besides Lakefield Farms, other "show farms" included:

Buehl Farms at Oxford. This wholesale hardware family from Detroit operated two dairy farms in the Oxford area.

Great Oaks Stock Farms at Rochester. Angus.

Hyup Farms at Birmingham. Holsteins.

Meadow Brook Farms at Rochester. This farm was owned by Mrs. Wilson, the former Mrs. John Dodge of auto fame. Belgian horses, beef cattle and hogs.

Wildwood Farms at Lake Orion, owned by William E. Scripps of the *Detroit News*. Guernsey dairy cattle, Angus beef cattle, Duroc hogs, Shropshire sheep and Belgian horses.

Windrow Farms at Birmingham. Ayrshires.

THE OTHER SIDE. The second story hay storage area of the barn that fronts on Clarkston Road has been remodeled into a beautiful living area. The old bull pens show the wear and tear of not having animals around for more than 40 years.

walked across it. It was like trying to walk on jelly. But your grandmother had no problem and often harvested berries from this land.

"Lambert and Webber also had some rocky land up in the hills behind the lake which was only good for sheep and cattle pasture," recalls Dad. "That ground is now covered with $250,000 homes.

"They also owned land on Clarkston Road around the Webber School. In fact, they gave that land to the school district to build a school for area residents and farm children which replaced the Shanghi school on Clarkston Road.

"Plus they owned all of the land along both sides of Baldwin Road except for what our family owned."

Dad remembers they built the fancy house on the lake between 1920 and 1925. Lambert's wife had died and he lived with his daughter, Peggy, who was Webber's wife.

"A chauffeur used to bring them out to the farm on weekends from Detroit," says Dad.

The Dennis Farm

Dennis Lake was named after the Dennis family who had a

THE BIG HOUSE. Located across the road from the dairy barns, this house was home for the Len Logan family and also for a number of single people who worked on the farm. Today, it serves as the office for the land development company.

farm at one end of the lake. However, Grandma Norah knew it as Lowry Lake, named after the farm family that lived there before the Dennis family bought them out.

"The Dennis brothers used to rent boats on the lake," says Dad. "They used to get some shady characters out there to go fishing and to rent the boats. There were three Dennis brothers and they were hard of hearing. You could hardly talk to them."

Lambert and Webber really wanted to control the entire lake. They kept upping the price for the land and the Dennis brothers finally sold out.

"Even your grandmother and grandfather Lessiter had 72 acres which they sold to Webber in the mid 1920s at an exorbitant price," says Dad. "The broker pestered my folks and it seemed like he came to see them daily.

"Your grandmother didn't want to sell the field west of the lake, which was really stony and where we raised potatoes. Your Grandpa Frank finally talked her into it as $500 per acre was hard to pass up in those days.

"The 40 acres they sold on the east side of the lake which came up to Baldwin Road probably contained the best soil of any field on our entire farm. It was valuable land and later was used to raise potatoes for the Lakefield operation."

What I Remember...

FOLLOWING ARE A FEW impressions of the Lakefield Farms operation as seen by a Lessiter youngster growing up down the road in the 1940s and 1950s and from stories told by Dad.

1 Even though our family members weren't actually part of the Lakefield Farms crew, we were always invited to their annual Christmas party. Everyone always treated us like part of their extended family.

2 There had always been some hassle about the sale of a segment of the Lessiter property in the 1920s and the question dealt with who owned the two-acre apple orchard located south of the big Lessiter house.

Even though the Lakefield Farms crew pruned and sprayed the trees and harvested the apples from this orchard, Grandma Norah always believed it was still our property.

Maybe she had a valid point. When the lawyers got involved in selling the property and clearing the title to the land in the mid-1950s, they needed her to sign off on some legalities in the paperwork for the property transaction completed 30 years earlier.

3 Only two women ever lived in the Foster farmhouse which was built at Lakefield Farms in 1924.

O.F. Foster's first wife, Nora, died of cancer in 1938 and left him with five young children.

Florence Sutliff came to Lake Orion to teach school and became the wife of O.F. Foster in 1941. She helped raise the five Foster children—Ruth, Jean, Virginia, Catherine and Jack. They later had a son, Richard.

After the death of their spouses, Florence Foster and John Lessiter were married in 1983 and continued to live in this house until Florence's death in November, 1995.

"Florence really worked hard raising those five kids and having buyers stay for dinner on the spur of the moment," says Dad.

4 One of my parents' good friends was Jenny Chapin, who with her husband, Clare, played bridge in a six-couple group with Mom and Dad every second Saturday night for over 25 years.

Her dad, Wilbur Priddy, earlier worked at Lakefield Farms as the first farm manager. They lived in one of the farm's big houses and took in boarders who worked on the farm. Since her mother died at an early age, Jenny spent time cooking and washing for the farm crew.

5 Lakefield Farms often hired several Jamaicans who came north for the summer and worked in the fields, often pulling weeds in the potato field.

They lived in an old 15-foot square chicken house without heat and only a bare lightbulb and simple stove. It certainly didn't provide much in the way of attractive living conditions.

6 The farm's hog man, Jerry Peters, was going bald and was using a special tonic to attempt to get hair to grow on his head. I remember the threshing crew sprinkling the tonic all over a teenager's chest in the hope he would soon sport some hair.

7 Because of his asthma, O.F. spent a great deal of time working on the farm books in the office remodeled from a bedroom in the house built for his family in 1924.

In the middle of each afternoon, he would get in his car and slowly drive down Clarkston and Baldwin Roads, looking at the progress made by the farm crew during the day.

8 I remember the time Katie Mattson put her foot down. When one of their daughters was born, her husband, Kent, came up with this great name. Katie thought it was okay, until she somehow learned it was the name of a super cow stanchioned in the barn. As you might guess, she vetoed the idea and they decided on another name, Heidi.

9 The Fosters were a deeply religious family, which meant the farm crew didn't do anything but livestock chores on the Sabbath. Sure, there was an occasional exception, but even the threat of rain ruining a downed alfalfa crop wouldn't normally lead them to bale hay on Sunday.

10 I remember Kent Mattson talking about leaving the Lakefield dairy barns and joining the U.S. Marines. He headed for California's Camp Pendleton for boot camp.

As the recruits gathered for the first time, Marine sergeants told them President Franklin Delano Roosevelt had scheduled a trip to Camp Pendleton in eight weeks to review the troops.

Kent may have exaggerated, but he said all they did for eight weeks was practice marching for the President's visit.

When the President came to California for his visit, the spit-and-polish platoon did an exceptional job of demonstrating their marching skills. Then they were immediately shipped off to Iwo Jima, where those marching skills honed over eight weeks of Marine training didn't do them much good in the hand-to-hand combat that followed.

11 I never saw Mom get very mad, but she wasn't very happy when Bill Hess dropped a pair of baby bunnies on our front porch one Easter Sunday as presents for my sister and me. She figured those young rabbits would be a lot of work and was really annoyed that Bill did this.

12 I still remember looking south from the farm during a violent rainstorm and seeing clouds of black smoke rising in the air.

One of the Lakefield barns had been struck by lightning and ended up being a total loss. I immediately pedaled my bike down there and watched as several fire departments feverishly tried to save the other barns.

What I remember most was the old tile silo. The temperature got so high in those air-filled tiles that they would pop and pieces of ceramic tile would go flying in every direction.

13 The first year the Michigan legislature passed a bill authorizing Daylight Savings Time, O. F. didn't like the idea. He was convinced that setting the clocks ahead an hour in the spring would confuse the high-producing Holstein cows and lead to a drop in milk production.

So the decision was made that the farm crew would remain on Eastern Standard Time even though everyone else in the state set their clocks ahead one hour.

What a mess! Employees showed up an hour ahead of schedule, some on time and others got to work an hour late. Kids were missing the school bus or were outside a full hour early.

Within a week, the farm switched to daylight savings time like everyone else in Michigan. And the cows never missed a milk-producing beat.

14 A teenager who worked on the farm during the summer, Bill Mervyn, carried the unofficial title of water commissioner. Actually, he was in charge of moving irrigation pipes and constantly changing sprinkler nozzles on the 40 acres of potatoes grown each year.

He spent his working hours traipsing among potato rows in hip boots and was always soaking wet during the hot summer weather.

15 Four Lakefield Farms couples and Mom and Dad played "500" each month for years. They eventually grew into a bunch of serious card-players that got together every month for nearly 40 years!

Kent Mattson hated to lose at anything and that included cards. He always tried to get Mom to be his card-playing partner because he knew that would increase his chances of winning.

16 Many nights in a darkened kitchen, Bill Hess showed me how to develop black-and-white film and make prints.

Along with the interest my mother showed in helping me develop as a writer, Bill's help turned me on to photography, which was a perfect foundation for a magazine editing career.

17 Lakefield Farms was known in livestock circles throughout North America for developing an excellent herd of Holsteins and an outstanding flock of Oxford and Hampshire sheep which were shown at the county and state black and white shows, the Michigan State Fair and the Saginaw County Fair. They may also have exhibited Duroc hogs in earlier years.

18 They kept a number of Belgian horses for field work until tractors really took command.

Two of Vic Kramer's prized Lakefield Farms Belgian horses, who were named Scout and Silver from Lone Ranger fame, eventually ended up working at our place in the early 1950s.

19 Lakefield dairy herdsman Kent Mattson and sheep man Bill Hess were good friends, but they liked to tease each other. At the time of the farm's dispersal in 1956, one of Mattson's best lines was the fact that all of the sheep put together didn't bring as much money as the price received for the top Holstein cow in the dispersal.

20 During the first day of the Oxford sheep flock sale in 1956, an Ohio buyer placed the top bid on 40 ewes and rams.

THE OWNER'S PLACE. The Lambert and Webber weekend home was built on the lake in the mid-1920s. A chauffeur used to drive the family out each weekend from their home in Detroit. Remodeled a number of times, it's still a very nice place to live.

When it came time to pay for the animals at the close of the sale, he told the auctioneer he was only bidding to make the sale more successful and wasn't in any position to pay for the sheep.

All of a sudden, the farm was stuck with 40 unpaid ewes and rams. But they lucked out.

As the Hampshire sheep sale was about to get underway the next morning, an Oxford breeder said he got confused and showed up a day late for the event. He wished he'd been on hand to buy some animals the previous day.

Quickly, a sale of the remaining 40 Oxford rams and ewes made everyone happy.

21 In early days on the farm, we used to sell milk to a creamery in Pontiac. Starting in 1934, the farm's milk went to the processing facility at Lakefield.

All of this processed milk was

THE CASTLE. Known locally by this name, this is an example of the expensive homes that have replaced corn, potatoes and forage at Lakefield Farms.

sold to Harper Hospital and the J.L. Hudson department store in downtown Detroit where Oscar Webber was the general manager. In later years, the milk was shipped in 5-gallon cans with rubber hoses and spigots which would be used in milk dispensers found in the hospital and department store dining rooms.

For years, Vic Kramer made a six-day-a-week, 90-mile trip to and from Detroit with the milk truck. This usually meant a mighty long day fighting morning rush hour traffic down Woodward Ave. and then waiting in the Hudson store unloading area while huge semi-trucks unloaded appliances, clothing, toys, furniture and other products.

"Selling our milk to Lakefield was a super opportunity and maybe one we didn't appreciate as much as we should have until they went out of business. Suddenly, we were selling our milk to a creamery in Pontiac and it was never the same," says Dad.

"We never paid any trucking or got socked with any quality discounts for the milk sold to Lakefield. They picked up the milk every day at no charge. It was a good deal."

22 One afternoon when Vic Kramer was pulling back out onto Baldwin Road after picking up our milk, his truck got smacked by a car.

Dad, hired man Harry Robertson and I heard this terrible collision and ran out to see what had happened. Vic was okay and no milk had been spilled, but the closed milk box was sitting on the road with the truck chassis in the center of the road next to it.

It took quite a bit of work on the part of the hastily assembled farm crew to lift the box back onto the frame.

23 Bill Hess got me started with a 4-H sheep project. He and Dad helped me pick out lambs for the Detroit Junior Livestock Show for several years.

Bill also sold me a half-dozen Hampshire ewes which I later sold to help finance my college education.

24 With the continued growth of homes in our area, roaming dogs were a serious problem for sheep producers. In fact, Dad got out of the sheep business in the mid-1940s because dogs chasing sheep were becoming a major problem.

You could talk until you were blue in the face with neighbors who would deny their dogs were left loose and certainly were not the cause of any sheep killed during the latest nighttime attack.

It got so bad that the Lakefield crew would shoot any dogs they found chasing sheep. The next morning they would bury the dogs before anyone realized their animals were missing.

The idea worked. Before long, neighbors were keeping a closer eye on their pets and not letting them stray at night.

25 One of the farm crew members got in serious trouble when he shot a sheep-chasing dog in a neighbor's yard. The only witness was a 5-year-old boy and the judge decided the youngster was too young to testify against the worker.

26 Thanks to Bill Hess and Kent Mattson, I spent several years as a teenager at the Michigan State Fair working with them in the sheep and dairy barns. It was a great experience to help load the trucks and head off for the big show in Detroit.

Those trips taught me a great deal about feeding and preparing livestock for the show ring.

27 One time I had a yearling Holstein heifer I was getting ready for the summer 4-H Club county show. Since she was due to calve shortly, Dad didn't want to put her down in a rope sling to trim her hooves.

Kent Mattson said he'd trim her in the stocks they used when trimming the Lakefield herd. So I walked the heifer a mile to the dairy barn, had her feet trimmed and walked her a mile back home. After walking two miles on the asphalt, her hooves were in perfect shape. And both of us were tired.

28 I left early one morning with Bill Hess for central Indiana on a trip which had a two-fold purpose. We had a half-dozen Hampshire ewes in the pickup truck to deliver to an Indiana sheep breeder.

But that wasn't the real purpose of this trip nor was it why the trip had to be made this particular day. After delivering the sheep to the Indiana farmer, we drove 35 miles to the farm of well-known Holstein breeder Harry Rosenbury.

Knowing in advance the approximate time of our arrival, he handed Bill a canister of freshly collected bull semen from one of his bulls and we drove practically nonstop all night back to Lakefield Farms.

The timing was right to breed one of the famous Delight family cows to this specific bull. This was before the days of frozen semen, so time was of the essence in breeding this Michigan based cow with refrigerated fresh semen collected only a few hours earlier from a bull in an Indiana bull pen.

29 It was a rather modern grain, feed, apple and potato storage and handling center when built, but the two-story, banked-hill, concrete-block, painted white structure was called the "White Elephant" by the farm crew.

"Actually, the building was part of the downfall of the health of O.F.," says Dad. "He suffered from severe asthma. One time he told one of the farm workers not to dump a load of grain until he was out of the building. For some reason, the worker did it anyway. O.F. had a very bad asthma attack from the resulting grain dust and was hardly the same after that."

30 O.F. and other members of the Lakefield team were always generous in allowing Dad to borrow farm equipment.

One example was a crop-dusting rig which they had permanently mounted on an International A tractor for controlling insects on 40 acres of potatoes. Borrowing this unit made it very easy for us to keep the insects under control on our few acres of spuds.

31 Len Logan was the farm's field boss and chief mechanic. I still remember the time he came up with the idea to save time by rigging the threshing machine blower directly over the hay baler.

The concept was that the straw would be baled as the wheat was threshed. This would mean the crew wouldn't have to build a straw stack out in the field and then later have to pitch all the straw into the baler.

The idea sounded great, but it actually turned out to be a costly mistake. The baler caught fire from all of the dust and chaff floating through the air. Quickly, the crew pulled the threshing machine away and watched the baler go up in flames.

32 Lakefield bought one of the first high-capacity silage choppers to come on the market in the late 1940s. They ran stake trucks alongside to take on big loads of silage and hauled the crop as far as 1 1/2 miles to silos at the dairy and sheep barns.

After coming home from school on fall afternoons, I loved to ride in the trucks with the truck drivers.

I'd ride along as the crew maneuvered the trucks in and out of the rows next to the chopper, watch all of the silage being blown into the trucks and enjoy traveling down the highway with huge loads of fresh green silage heading for the silos.

Today's monster choppers make those rigs look like toys. And nobody today chops silage as green and as wet as we did 40 years ago.

33 One of the wildest tests of farm machinery I ever saw happened one fall. After the field had been plowed, engineers from the Ford Tractor Company rented the field.

Night and day for five days, they ran hay rakes over the plowed fields to test the endurance of rake teeth!

Everyone thought the Ford engineers were crazy. Travelers on the way to work in General Motors plants in Pontiac who didn't know what was going on probably figured the Lakefield Farms crew had lost their minds.

Imagine what they thought of the farm crew using a hay rake to smooth out the ground in a plowed field!

One Of America's Best Dairy Herds

OVER A 42-YEAR SPAN, three men—Oscar Webber, O.F. Foster and Kent Mattson—put together one of America's great Holstein herds of all time. They did it at Lakefield Farms, the Clarkston, Michigan, farming operation located right next to the Lessiter family farm.

Oscar Webber

A highly successful Detroit businessman, Webber and his father-in-law, John Lambert, started to put together the farming operation in 1914.

Webber was well known as a Detroit civic leader, and was always keenly interested in agriculture and the improvement of purebred livestock.

O.F. Foster

A University of Illinois graduate from southern Illinois, O. F. Foster came to Lakefield Farms in 1924 as the hog man. A few years later, he became the farm's manager and stayed for 32 years.

"His livestock program, together with his farm crops, resulted in his advice being sought in every phase of agriculture," wrote C.B. Smith, sales manager in the Lakefield Holstein dispersal sale catalog in 1956.

"A keen judge of a dairy cow,

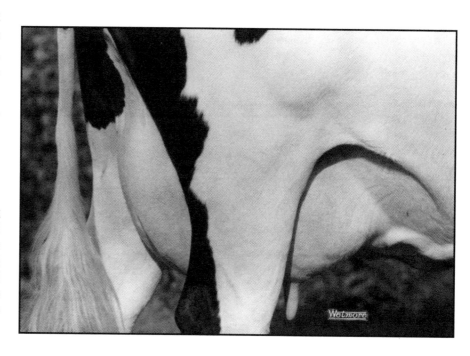

THE MONEY MACHINE. High milk production is what dairying is all about. This is the udder on one of the herd's best cows, Lakefield Fobes Delight. She captured best udder honors at both the 1953 and 1954 editions of the Michigan State Fair.

MICHIGAN STATE FAIR. For years, Lakefield Farms took home the lion's share of the Holstein show ring awards at this annual event held in Detroit.

he always insisted upon both type and production as shown by the sensational records and year after year Premier Breeder at the Michigan State Fair."

Kent Mattson

Kent Mattson went directly from school in the western part of Michigan's Upper Peninsula to Lakefield Farms in 1940 and shortly afterward entered the U.S. Marines.

Returning after World War II, he spent 16 years working with this great herd of Holsteins before moving on to working with the black and white cows at Carnation Farms near Carnation, Washington, and the Apache Ranch at Lapeer, Michigan.

"A great deal of the success of the outstanding records is due to his untiring effort and constant care of the cows in his charge," wrote C.B. Smith at the time of the herd's dispersal in the spring of 1956.

At the time of the sale, this was the only herd in America ever to develop three twice-a-day milked cows each having over three 1,100-pound butterfat records.

In earlier years, the farm also

BEST FOUR DAUGHTERS. These daughters of Gold Medal Sire Lakefield Winterthar Victor Fobes were selected as top Get of Sire at the 1954 Michigan State Fair.

OUTSTANDING BREEDING. Many outstanding cows and bulls were developed in the Lakefield Farms herd. When classified in the Spring of 1955, the herd included 8 Excellent, 20 Very Good, 24 Good Plus and 13 Good Holsteins.

employed many other outstanding cattlemen who went on to later successes. One was Frank Case, herdsman at Lakefield Farms from 1938 to 1941.

In the summer of 1937 on a typical windy, dusty day in Fillmore County, Nebraska, Case received a telegram from Foster asking if he would be interested in the Lakefield herdsman job.

He had been recommended by national Holstein fieldman Ernie Clark, who knew of Case's work in the Nebraska Holstein Association and the well-known herd he had developed on the home farm.

So in January of 1938, Frank and Marie Case and their family headed for Michigan in a 1936 Chevrolet, pulling a homemade trailer carrying their belongings. When they arrived, they immediately sold their homemade trailer to raise needed cash.

The Case family spent almost four fruitful years at Lakefield Farms. During this time, the farm gained the honor of having the highest herd average for butterfat among U.S. herds milking more than 50 cows twice a day.

Case always remembered December 7, 1941, (Pearl Harbor Day) for several reasons. Not only was it the official start of World War II, but the day he became farm manager for the 345-acre Kyland Farms at Oconomowoc, Wisconsin.

Owned by Milwaukee businessman, W.D. Kyle, Jr., the old farm buildings were soon replaced with a complete new setup and the goal of developing one of the country's outstanding Holstein herds was established.

No Holstein herd at that time had ever developed a line and inbreeding program as successfully as the one used by Kyland Farms.

In later years, Case gave much of the credit for his extensive knowledge of farm management to the years spent working with Foster.

The Lakefield herd was a member of the Holstein Association's Herd Improvement Registry testing program for 27 years. Over a 16-year period from 1939 to 1955, 33 cows averaged 14,631 pounds of milk and 533 pounds of butterfat.

By 1956, the herd included nine Excellent, 26 Very Good, 22 Good Plus and four Good Holsteins.

The Dispersal

As Oscar Webber was growing older, there was little interest in the farm among the next generations and the family's estate plans became a concern. The decision was made in the mid-1950s to sell the livestock and the acreage at Lakefield Farms.

Held in a tent on the farm grounds on May 18 and 19, 1956, the Lakefield Farms dispersal closed with a bang heard 'round the world. Over 1,000 people were in attendance with buyers from 14 states, Colombia and Guatemala.

In two days, 151 Holsteins in the 42-year-old herd were dispersed for an average price of $1,248.18 per head.

This was a sale price un-

THE $13,000 COW. At the Lakefield Farms dispersal, Lakefield Fobes Delight topped the sale with a $13,000 bid from Carnation Farms at Carnation, Washington. At the halter of this famous cow is Lakefield herdsman Kent Mattson. Next to him is Russ Pfeiffer, long-time Carnation Farms manager. Next is O.F. Foster, manager of Lakefield Farms.

RECORD SETTING AVERAGE. In two days, 151 Holsteins in the 42-year-old Lakefield herd were sold for an average of $1,248.18. This was an average unequaled by a registered Holstein herd of this size in more than a dozen years.

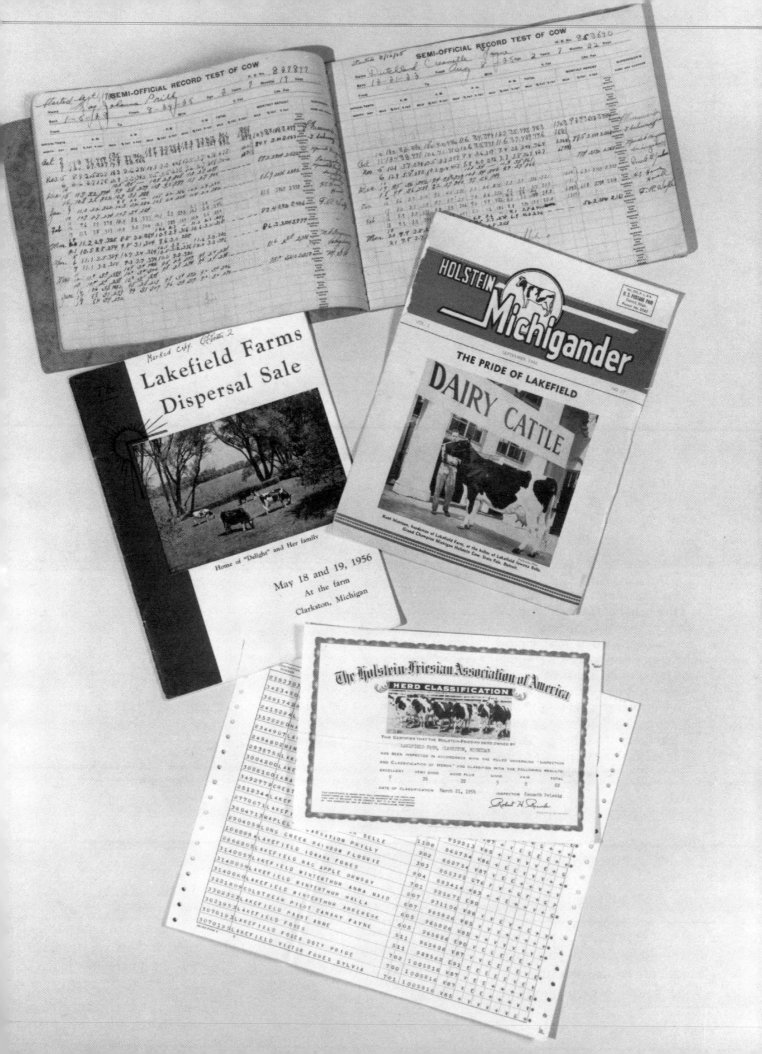

DELIGHT AND HER FAMILY. Lakefield Farms was known as the home of "Delight" and her family. Shown in the foreground is the family's patriach, Minnow Creek Eden Delight. At right in the foreground is the dispersal's top seller, Lakefield Fobes Delight, which sold for $13,000. In the background are two other members of the Delight family grazing along the shore of the farm's Dennis Lake.

equaled by a dispersal since the Creston and Butterfly herd sales of 1947 or by a herd this large since the Dunloggin Dispersal in 1943.

The top price of $13,000 for Lakefield Fobes Delight paid by Carnation Farms of Carnation, Washington, was unmatched by any female since Dunloggin Mistress La Princess went through the 1947 Curtis Candy Classic at $23,500. The 2-year-old son of Fobes Delight brought $10,000 and 50 other animals sold for $1,000 to $8,500.

The old family patriarch, the 14-year-old Minnow Creek Eden Delight, sold for $8,500 to J.E. Taylor of Meadow Farm Dairy in Orange, Virginia. She had just wrapped up her twelfth lactation and her fifth lactation with over 1,000 pounds of butterfat on twice-a-day milking—the only cow in the world to have done so at the time.

Her lifetime records in the sale catalog showed production of 192,352 pounds of milk through a dozen lactations.

At the time of the sale, she was bred to her $10,000 grandson, Lakefield Fond Delight Fobes, a 3-year old bull purchased at the dispersal by W.W. Sherman of Bloomfield, Connecticut.

The nine members of the Delight family brought a total of $51,200.

Hire The Best!

Top people made the Lakefield cattle operation what it was during their successful 42-year reign. And farm manager O.F. Foster was no exception.

"He was the most fair-minded and most respected man I ever

knew," said veteran show-ring competitor Everett Miller some 27 years after the Lakefield dispersal. Miller had served as herdsman and farm manager at Hyup Farms in Birmingham, Michigan, and went head-to-head many times in the show ring with the outstanding Lakefield Holsteins at county, regional and state shows for a number of years.

"When he was managing Lakefield Farms, O.F. had money to spend, but he seldom spent it at big sales," added Miller. "You'd see him at farm sales looking for diamonds in the rough. Being a great cowman, he knew how to find them.

"He was just as good with sheep and hogs. A real livestock man."

The Special $500 Cow

One of the less noteworthy purchases at the Lakefield Farms dispersal was the $500 paid for a 5-year-old cow, Lakefield Fobes Bonheur, by Dad.

This cow was purchased to provide my 9-year-old sister, Janet, with a 4-H Club calf.

Months later, the big black cow came through with the wanted female calf which went on to win a number of awards.

In fact, the resulting calf serves as the model for the "Blossom calf" featured on pages 209 to 214 in this book.

THREE GENERATIONS. Top to bottom at left, Minnow Creek Eden Delight, Lakefield Fobes Delight and Lakefield Fond Delight Fobes. Winning senior yearling honors at the 1955 Michigan State Fair, the bull was a son of the Fobes cow and a grandson of the famous Delight.

History Hung On A Thread

A GREAT CHAPTER of Holstein breed history hung for a brief moment back in 1938 on a slender thread of chance!

There would have been no Minnow Creek Eden Delight family at Lakefield Farms if Glenn C. Wilson of Wanatah, Indiana, had not come down to Harry Rosenbury's farm at Rochester, Indiana, on a certain day that year looking for a bull to put to use in his herd.

Rosenbury took him over to a neighbor's farm, the operation of James W. Vanlue, to see a 3-year-old sire named Perfection Ormsby King (POK) that Vanlue was finished using. Unfortunately, they discovered the bull had just been shipped to the stockyards in Indianapolis.

The two men were impressed with the bull's daughters and went home extremely disappointed that they were unable to purchase the bull.

Two days later, quite by accident, Rosenbury learned the bull had not yet been slaughtered. He hurried to the Indianapolis stockyards, found the bull had already been sold to a packing company for a nickel a pound. By paying them a nice profit, he was able to take the bull home for a total cost of $100.60.

Not long after this, Rosenbury went back to Vanlue and purchased an open yearling daughter of POK for $62.50. He bred her to her own sire, getting an inbred black heifer calf named POK Mandy.

The Mandy heifer soon became the special pet of the entire Rosenbury family. When she came of breeding age, she was bred artifically to the great young Maytag-bred bull, Posch Ormsby Fobes Eden.

Minnow Creek Eden Delight was the result of this mating. Soon bred back to Eden, Mandy next produced Delight's full sister, Minnow Creek Eden Repeat.

The rest of the story is well known, but it is interesting to note that Rosenbury's capital gain on his $163.10 investment in Mandy's sire and dam, plus the cost of three breeding fees, amounted to 4,400 percent after he had sold Repeat for $3,300 in the first Wolverine Classic and Delight and her young Roamer son for $3,600 and $3,100, respectively, in the 1949 Wolverine Classic—a total of $10,000.

Lakefield's $53,200 return from their $3,600 investment in Delight (including the $2,000 son of "Fobes Delight" sold in the Pan American sale in the fall of 1955) was 1,475 percent.

—*June 2, 1956, Holstein World*

MINNOW CREEK EDEN DELIGHT. Just over 13 years old at the time of the Lakefield Farms dispersal in 1955, this extraordinary cow had produced 1,151 pounds of butterfat and her fifth twice-a-day milking record of over 1,000 pounds.

When Ford Didn't Have The Bucks

IT WAS THE summer of 1902 when a gentleman farmer from Detroit stopped at the Lessiter family farm to look over a small group of young Shorthorn bulls advertised for sale.

My grandfather Frank and his brother, Floyd, ("Uncle Jay") had heard of the "Detroiter" before—he was known pretty much as a "crackpot" because of some of his crazy ideas.

Yet he later became one of America's most successful and richest businessmen.

Horseless Carriage Man

The gentleman farmer was Henry Ford and his visit to our family farm came less than a year before he began rolling his "horseless carriages" out of an old converted wagon factory in Detroit.

The amusing details of how Ford happened to come to our farm on his bull buying trip is a story that has been passed down through five generations of our family over the past 90-plus years.

As I heard the story from my Dad and Grandmother Norah, Ford arrived at our Lake Orion farm after a 35-mile journey north from Detroit in what was a

JUST A YEAR AFTER Henry Ford visited the farm, he turned out this very first model of his 1903 Ford Runabout.

BEFORE HE EVER WAS an automobile manufacturer, Henry Ford was a gentleman farmer near Dearborn, Mich.

forerunner of the company's most famous car—the celebrated Model A. He was still experimenting with the machine at that stage and he came with a mechanic who did the driving.

Problem Prone Machine

There apparently were many problems with the car during the half-day jaunt. After arriving at the farm, the mechanic spent all his time tinkering with the cantankerous horseless carriage while Ford looked over the Shorthorn cattle.

As Dad tells it, neither Grandfather Frank nor Uncle Floyd were very impressed with what they saw—which was one of the very first cars either of them had ever seen up close.

Anyway, while the mechanic worked on the car, Ford and my Grandfather and Uncle Floyd walked into the pasture to look over a group of young Shorthorn bulls. The family herd had been winning quite a few blue ribbons at various fairs around the state which was what attracted Ford to our cattle breeding program.

A Real Standoff

Once in the pasture, Ford finally selected a young yearling Shorthorn bull he liked. After the three men did some serious haggling, they finally agreed on a price for the bull. Then the fireworks really got started!

Ford wanted to write a check just large enough to cover a down payment and arrange an installment contract for the remainder. My grandfather and Uncle Floyd would have none of that—they were a little wary of Ford in the first place, and insisted on cash for the bull!

Ford wouldn't relent—trying to launch his car company at the time, he was short of cash and kept insisting on negotiating an installment deal on the bull.

When the three men couldn't agree on financial terms, the bull kept grazing in our pasture.

Ford stomped back to the car and told the mechanic to finish adjusting the car's engine, because they were leaving! They soon took off in a huff. That was the first and last time Henry Ford ever visited our farm.

Grandpa and Uncle Floyd felt they'd done the right thing—with the problems they'd seen

---

"When Ford couldn't come up with the cash, Grandpa wouldn't sell him the bull..."

Ford have with the experimental car he'd brought to our farm, they certainly didn't think they should risk the payments for a good Shorthorn bull on that sputtering machine's future.

Secondly, the earlier genera-

HENRY FORD operated a general farm which also included a herd of valuable Shorthorn cattle, the reason he visited the Lessiter family farm back in 1902.

tion of our conservative family wasn't about to take a chance on anybody who actually believed these horseless carriages would ever totally replace horses.

Ford never mentioned the

ALWAYS A FARMER AT HEART, Ford soon found himself involved extensively in the tractor business. One of his first Ford tractors is shown here tilling the soil on the Ford farm.

329

possibility of offering stock in his Ford Motor Co. for the bull during the visit to our farm. But he later asked a friend of our family to think about investing $500 in his proposed auto company.

The friend considered it for awhile, then turned Ford down and was not among the original 11 investors who eventually raised $28,000 among themselves to form the Ford Motor Co. in 1903.

A Weak Beginning

By the time Chicago dentist Ernst Pfenning bought Ford's first model L car in July of 1903, the company's capital had dwindled to $223.65. But you and everyone else know how the car quickly proved to be a big hit, and how Ford's company soon grew to become one of this country's leading businesses.

Our family has often wondered what that neighboring farmer's investment of $500 in 1903 would be worth today, but we all agree it would surely have a value of millions of dollars.

At any rate, as Ford became much more successful over the

"Henry Ford took off in a huff, the first and last time he ever set foot on our farm.."

years, the five generations of our family have enjoyed some good laughs about that summer day in 1902, when Henry Ford's check and credit just weren't good enough for the Lessiter family!

HAULING BIG LOADS OF HAY at the Ford family farm meant relying on the auto company's products for needed power. Note the street car tracks and brick roads in the area around the farm.

When Dad Served On The School Board

SERVING ON THE local school board is normally a thankless job. Since most people are unhappy with some little aspect of education, they tend to take it out on the board members.

Dad served a long stint as President of the Lake Orion School Board in the late 1940s and early 1950s while I was growing up. While his being President of the board probably kept me out of serious trouble with teachers and principals, there were drawbacks.

Yet by the time my sister, Janet, who is seven years younger than me, started school, Dad had retired from the school board and those drawbacks had disappeared.

That just goes to show you once again the youngest child always gets all the breaks.

Starting with kindergarten and continuing through high school, the kids living in the farm's tenant house and I had to walk a half-mile to the bus stop.

You Can Walk!

A couple of years after Dad became a member of the board, school officials tried to change the route so the bus would stop in front of our house.

Dad put his foot down and told school officials "no"—that people would think he was using his influence as School Board President to make it easier for his kid to get to school. So I continued to walk the half-mile to the bus—and yes, it was uphill both ways!

But wouldn't you know it! Soon after Dad retired from the school board and I graduated from high school, the bus started to stop at the house to pick up my sister when she switched to Webber School.

Snowy Weather

We usually had a couple of snow days during the rough Michigan winters that meant the schools unofficially closed down for the day.

About half the teachers would make it to school on these snowy or icy days, the buses would be canceled and only a half-dozen town kids would show up for classes.

But there was always one country kid who somehow showed up for classes even though the buses had been canceled for the day—me!

As President of the School Board, Dad's philosophy was that he and his kids had to set the example for the community. On these days, this meant I had to get ready for school and then he and I would set out in the pickup truck over the icy or snowy roads for the hazardous 5 1/2-mile trip to school. This also meant he had to come back in the late afternoon to pick me up.

Nothing was ever learned on these "snow days" because the teachers knew they'd have to repeat any lessons the next day for the hundreds of kids who enjoyed their day off. So all we ever did was play games on these snowy days.

You guessed it! Just as soon as Dad got off the board, my sister never had to go to school on these really bad winter days.

Continuing Education

While I was in high school and a 4-H Club member, Mom and Dad usually made a late February trip to Michigan State University for the mid-winter Farmer's Week. This was a unique educational program for farmers featuring hundreds of speakers on every possible agricultural subject and meetings of farm groups that had been held every winter on the East Lansing campus for decades.

Many of my 4-H Club friends attended one or more days each winter. But Dad wouldn't budge when it came to me wanting to go for just one day.

He felt seeing the President of the school board allow his kid to miss school would be setting a bad example for the community. He maintained I needed to stay in class and study hard to earn better grades.

Even when a family friend who was the school principal, Vena Kirkpatrick, told him it would be a good educational experience for me, he still wasn't convinced. She told him to quit being so stubborn and to take me to Farmer's Week, but it still didn't happen.

I don't know if my sister ever took a day off to go to Farmer's Week or not, but if she didn't, it's only because she didn't ask. But I do have to admit she got all A's in school...something you didn't see on my report cards.

Livestock Shows

For nearly all of the 4-H kids raising fat lambs, barrows and steers, the annual early December Detroit Junior Livestock Show was a three-day affair. For me, it was only 2 1/2 days!

The schedule called for checking in and weighing animals early on Tuesday, the judging was held on Wednesday and the livestock auction on Thursday.

A number of school friends showed steers and they'd load up early on a Tuesday morning and head for the Michigan State Fairgrounds in Detroit. They would arrive about 10 a.m. and have the rest of the day to get their animals settled, weighed, cleaned up and have some fun.

Me? I'd still be in the classroom until the lunch break at noon. Only then would Dad pick me up at school and we'd make the 35-mile trip to the Detroit show. My lambs were usually among the last of the livestock to be weighed on Tuesday afternoon.

It made me mad that I missed a half-day of fun. But it never did any good.

Made Some Trips

Yet I did participate in a number of exciting 4-H club events around the country while in high school. Dad did let me go on these trips, but it's probably because he was off the school board by that time.

So the next time your Mom or Dad thinks about running for the school board, you kids might want to zero in on their educational philosophies as it applies to you being the perfect student or setting an example for the whole town.

From experience, I know it's something you need to find out— now!

NO FAVORITISM. When Dad was on the school board, the kids walked a half-mile to the bus. When Dad got off the board, the bus stopped in front of the house.

Everything I Ever Learned...

AS A NEW kindergartener on Monday afternoon, September 11, 1944, I walked in the front door of what was then the only school in Lake Orion.

Some 13 years later on June 5, 1957, I walked out of that building with a high school diploma in my hand.

Built in 1927, this building was school for me for 2,340 days—the only school I ever attended during my grade school, junior high and high school days. In fact, I was part of the last class of students to spend all 13 years in the same school building.

While I was in school, the district built a new Blanche Sims grade school and a new high school which was completed during the summer of 1957. Each time, the class immediately following ours had the opportunity to spend a year in both new schools.

Going Back

They say you can never really go back to your youth, but that's not the case when it came to reliving all of the fun and learning that took place during my school days.

Some 38 years after receiving my diploma, Dad and I walked through the school one August morning. While Dad had never attended this school, it was a homecoming for him too, since

THE SCHOOL. The only school in town for many years, it is now used mainly for adult education and pre-school classes.

he had served as a member of the Lake Orion School Board for many years.

Dad and I started our school tour on the lower level where my kindergarten school career had gotten underway, a room that was later converted into the school cafeteria.

I recalled many memories of the years spent in the old school. As you will see, my academic learning wasn't what I remembered most about the 13 years spent here.

Kindergarten: Mom would drive me the half-mile to the bus stop and wait patiently for the afternoon session bus. Like most kindergarteners, I carried a piece of carpet to lay down on for a quick midafternoon nap or at least a few quiet, restful minutes for the teacher, Miss Smith.

First Grade: We moved upstairs to Ms. Detwlier's room where I learned to read. She let me and two other students go into the nearby cloak room and read at our own pace since reading came easy to us.

Second Grade: Longtime family friend Vena Kirkpatrick who would later go on to be the principal at the Blanche Sims school and assistant superintendent, was my second grade teacher. She had seen my crazy antics many times before, so she was prepared. She was a challenging teacher and taught me plenty. I was always leary of what she might be telling Mom about me, although I don't think it ever kept me from being wild!

SEVENTH GRADE MATH. This was the classroom for junior high math classes, one now used by pre-schoolers.

Third Grade: Miss Smith, the sister of my kindergarten teacher, put up with me in third grade. This was the year I slammed one of the glass doors on a classmate and watched in horror as blood spewed all over the place from his badly cut hand.

This led to installation of safety guards on the school's glass doors the next day.

Fourth Grade: Mrs. Perkins was certainly no disciplinarian and had her hands full. Another

---

"Actually, the fourth grade lipstick incident wasn't any big deal..."

---

boy and I were so unruly that he was shifted to another class partway through the year. I always suspected he was moved instead of me since Dad was president of the school board.

Another boy and I got caught smearing lipstick on the faces of girls during recess. Mrs. Perkins called the two of us up in front of the class and put lipstick on the two of us. It wasn't any big deal to me, but the other boy cried in embarrassment.

Fifth Grade: Mrs. Fiebelkorn was our teacher. She was a strict disciplinarian and made everyone toe the line.

One day I got in a fist fight with another student and Mrs. Fiebelkorn broke it up. That afternoon, we went to an assembly in the gym and halfway through the program, someone jabbed me in the ribs and told me to look across the gym at the stairway. Mom was standing there!

I was scared the teachers had called to tell her about the fight. Later on, I found out she didn't know anything about it—so I didn't bother to tell her, either.

Mrs. Fiebelkorn always called roll first thing in the morning and my good friend Larry Leach was right ahead of me in the alphabetical listing.

When she called his name, he would answer, "Here today."

Next, she would call my name and I'd quickly answer, "Gone tomorrow." Everyone in the room would laugh, but for some reason, I was always the one who got in trouble.

A few years later, Mrs. Fiebelkorn's son, Red, was a fighter pilot during the Korean War. He was shot down, listed as officially missing in action and finally declared dead.

The new high school basketball coach that year was Gaylord Townsend, who had been a player on the 1948 and 1949 University of Kentucky Wildcats national championship teams. He may have been tall and talented on the court, but he couldn't coach worth a darn.

During the school year, investigators came to the school to interview him about the Kentucky point shaving scandals, but he was found innocent.

The high school team didn't win a single game that year and he was replaced before the next fall rolled around.

He ended up working in downtown Pontiac for a jewelry store— which just happened to like being the sponsor of the City Basketball League Championship team every winter. I don't know how successful he was at selling jewelry, but his basketball talents gave the store a winning team.

Sixth Grade: Miss Zag was in her first year of teaching and had graduated the previous year from Ohio State University.

A real Buckeye fan, she and a friend went by train to Columbus for the end-of-the-season Ohio State and University of Michigan football game in November of 1950. That was the famous "snow bowl" game which Michigan won with a field goal being the only score in a real blizzard.

There was so much snow that Miss Zag didn't make it back for class on Monday since the snowbound trains couldn't run for several days.

As had been the case in fourth grade, I spent a great deal of time standing out in the hallway instead of being in class—at the teacher's request.

This got me in big trouble one day. I happened to be standing out in the hall when other students were released for band. Since I was a drummer, I went along with them to band class.

For some reason, Miss Zag thought I should have remained in the hallway instead of going to band class. And apparently she and the principal were shocked when she couldn't find me still in the hall.

Actually, my two worst years were in fourth and sixth grades. It seemed like I had my biggest problems with teachers who didn't control their students.

By comparison, I never had a problem in fifth grade since Mrs. Fiebelkorn would never stand for it.

Seventh Grade: This was the year my classmates and I moved up to the second floor of the school.

Another longtime family acquaintance, Mrs. Hemingway, was our homeroom teacher. She

---

Ketchup Sandwiches

A big treat after getting home from school was eating ketchup sandwiches.

The recipe isn't all that complicated: Take two slices of bread and pour on as much ketchup as you can.

Many years later, our grandson, Alex, relishes the same late afternoon after school snack.

BUSY STREET YEARS AGO. Since the school is no longer used for up to 13 grades of students, the street out front isn't jammed with traffic today.

was a little skeptical of my behavior and had probably heard bad things about me from some of the grade school teachers.

Mom and my Uncle Sherm had also had Mrs. Hemingway for English class. Because of some long-time class differences Mom had with this teacher, Uncle Sherm and I reasoned that was why we didn't always get an A in her class.

Eighth Grade: Mr. Doerr was our homeroom teacher. The high school basketball coach, Mr. Campbell, really turned the team around and they captured the district championship a year later. He also taught eighth grade science, but his mind was on basketball and our class had plenty of disciplinary problems. He started out putting students in the hall, but that didn't work when other teachers would see 10 boys having a good time out there.

Next, he came up with the idea that if you spoke out of turn, you had to spend the rest of the class time standing up. Each time you talked, you had to stand up for another day.

When that didn't work, he had the students stand a foot away from the wall, lean over and support themselves by placing their fingers on the wall for the entire class period.

It was no fun standing up, so I learned to keep my mouth shut and got to sit down before long.

BUSHMAN'S CAFE. In the late 1940s, many school kids ate lunch in a grill in this buiilding. Later, the school's hot lunch program put an end to eating here and playing pinball. The Bushmans later became restaurant owners in Oxford.

Avoid All Strangers

As a first grader, my sister Janet was cautioned by Mom not to ride with strangers.

"A neighbor, O.F. Foster, was really mad when Janet wouldn't get in the car with him at the bus stop and she walked the half-mile home instead," says Dad.

"I don't think she considered him a stranger, but for some reason she thought Mom didn't want her to ride with anyone."

Other classmates didn't fare so well. With 40 days of school remaining, some boys still needed to stand for 60-plus days.

Mr. Campbell did well with the basketball team and was a pretty good science teacher, but he sure had problems keeping everyone quiet in our class.

Because of an English class taught by Mrs. Hemingway, I learned a poem by heart after the school year was completed and I can still recite it perfectly today—more than 40 years later.

Each of her students had to be able to recite the poem, "Abo Ben Adem" during the year's last marking period. I thought it was a dumb assignment and didn't learn it. So Mom was pretty shocked when I brought my end-of-the year report card home that showed a low English grade for that marking period.

Mom was mad! I planned to go to 4-H Club Week in East Lansing in four weeks, but she told me I wasn't going until I learned that poem.

I pleaded with her and said the assignment had been stupid. But the night before I was to go to East Lansing, I recited every word to her.

Ninth Grade: We were finally in high school—big-time players in the game of education! Jim Hoag was our homeroom teacher. This was the first of many years he would spend in the Lake Orion school system as a teacher, football coach and ad-

ministrator. We spent most of the homeroom periods discussing sports.

Our class started ninth grade with 233 students and the number would be down to 99 by the time we graduated four years later. Besides the usual high school dropouts, another big reason for the drop in student numbers was the impact which the auto manufacturing situation had on many people in the Detroit, Pontiac and Flint areas.

Things were booming when our class started ninth grade and a number of families from the South moved here to enjoy the good wages to be earned by working in the Fisher Body, GMC Truck and Pontiac Motors auto plants. By the time we graduated in 1957, things were tough in the auto business and many laid-off workers had headed back South and taken their kids with them. Plus, many didn't like the cold Michigan winters.

I remember taking the career interest tests in ninth grade social studies class. Friends scored very high in one or two areas, giving them and the school

> *"We started with 233 students. Four years later, only 99 graduated..."*

counselors a pretty good idea as to what they wanted to be when they grew up.

The test results really scared me since I didn't score high or low in anything. Instead, I was pretty much average in interest in all of the job-related categories. It made me wonder what I was going to amount to later in life.

As it turned out, those tests and scores weren't worth the paper they were written on. And I guess ending up as a journalist proved I was interested—at least a little—in just about everything.

This was also the year of the "wild goose chase." This was the traditional Spring-time biology

THAT EMPTY FEELING. Going back to the school that almost seemed like home for 13 years, you sense a wave of emptiness. Quiet seems to have settled over the old building.

class trip to the Jack Minor Bird Sanctuary located at Point Peele, Ontario. As the teacher had us tell our parents, the idea was to study firsthand the migration habits of flocks of Canadian geese as they made their way north after their long Winter vacation.

There was only one thing wrong with this idea. By the time we visited the bird sanctuary in late April, there wasn't a goose to be seen. They'd all flown north a few weeks earlier.

So what did we do? We camped out and went crazy in the Canadian town of Leamington on Saturday night, lighting hundreds of firecrackers which could be easily purchased in Canada yet were not available back home.

Then on the way back to the Canadian campsite, Mr. Stoolmueller, our teacher, drove the bus under some low tree limbs and put a costly series of dents every two feet in the roof of the bus from front to back.

This was the last field trip to Point Peele which the biology class was allowed to make—an indicator of things to come when it came to making school trips later in our high school careers.

Tenth Grade: By now, Dad had been off the school board for a couple of years, but teachers still knew who I was. Another longterm family friend, Miss McGuffie, was the geometry, trigonometry and solid geometry teacher. Others in the family proved to be better math students for her.

The English teacher that year was Mr. Michael and along with Mom, who had been an English teacher herself, got me interested in journalism. He had apparently noticed some writing talent in me and really encouraged me in this area.

Interest in the high school newspaper was being rekindled and he was the advisor. He soon had me writing a feature for each issue and I found I really liked newspaper writing and the challenge of meeting deadlines.

This was also the year I took the most valuable class which I ever participated in during these 13 years in school. It was Mr. Crother's typing class and I worked my typing speed up to 60 words per minute by the end of the school year.

That typing class did more for

WINDING STAIRS. Located in the center of the school was a winding staircase used by students going to classes.

my career than anything else I have ever studied! Besides letting me get things down fast on paper, typing encouraged me to think and act at a fast pace— a big asset later in life.

Eleventh Grade: Mr. Egner was our American History teacher and did a pretty good job of relating all the interesting moments in history to us. But what I remember most about his second hour class had nothing to do with him or American History.

Instead, it had to do with a Lake Orion hockey team which played in a winter league in Pontiac. Even back then, reserving ice time was a problem.

The team often played league games late at night. This meant the players would be plenty tired by the time they got home.

They often showed up for school around 10 a.m. instead of 8 a.m. like the rest of us. Mr. Egner would be in the middle of a lecture about American history when several still bleary-eyed students would stumble into the classroom. I wonder how some ever passed their classes that winter as hockey was definitely their No. 1 priority.

Twelfth Grade: Our final year! Soon we would be free! I'd had a run-in with Mr. Leith, the band director, the previous spring. Since band had been switched to the last period of the day, it brought my drum-playing band career to a close.

Times were tough on the farm and Dad had to let the hired man go. I'd get up in the morning and help Dad with the chores, hurry to the house to take a bath and eat breakfast before driving the farm truck off to school.

I'd wrap up my classes by 12:30 p.m., eat with the third shift in the cafeteria and head home at 1 p.m. While my fellow classmates still had several hours of classes to go, I would be helping Dad with the farm work.

In December of that year, Mom's Aunt Belle died in Ontario and the rest of the family spent four days over there for the funeral.

I milked all the cows twice a day, did all the feeding and other farm work and still managed to make it to school on time each day. Those were pretty long days and I was sure glad to see Dad come home from Canada.

Even though I've been to the "Big Apple" many times since, I still recall our senior trip by train to New York City. Some of my classmates were pretty wild and got in trouble—serious trouble.

There were some under-age drinking problems. After toppling over in an elevator and being reported by the hotel staff,

WALKING THE HALLS. It's much different today than years ago when Dad was on the school board and hundreds of students sped down these halls hurrying from class to class.

Education Was A Must!

THE LESSITER FAMILY always believed in the value of a good education and it started with grade school. Several generations of the family attended

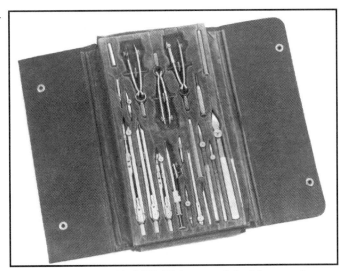

FAMILY DRAFTING SET. This set of drafting tools was passed down among three generations of Lessiter family members and was used in industrial arts classes.

the grade one through eight single-classroom Block School, located a mile northwest of the farm.

Later, Dad would go to high school in Oxford and live part of the time with Grandma Wiser who had moved off the farm to a home in Oxford.

Mom skipped a grade or two in grade school and graduated early. "We sat at double desks, saluted the flag, had prayer, ate the lunches we carried from home in our rooms and received lots of individual attention and love," recalled Mom. "Knuckles were occasionally rapped and spankings administered."

Mom recalled it took several elections before a bond issue could be passed and a new school could be built in 1927, just as she and eight classmates were graduating from eighth grade. Her class was the first to attend the new school for all four high school years. Among the frills offered for the first time in the new school were physical education, French, home economics and shop.

On the way home from school, Mom often stopped at the blacksmith shop behind the *Lake Orion Review* office or Hessler's Cider Mill on West Flint Street. A fascination during the winter was watching men cut ice on the lake, store the large ice slabs in ice houses across from the dam or watch ice being loaded into freight cars for the journey to Detroit.

Later, Mom went on to Central Michigan University in Mt. Pleasant and eventually taught English and supervised the award-winning debate squad for several years at Fowler, Michigan, before coming back to Lake Orion to get married.

The Fowler High School in those days had so few students that they did not field a football team and played baseball in both fall and spring.

OLD BLOCK SCHOOL. The photo at left compares the school house in 1917 with the house it became nearly a half century later.

SCHOOL DAYS. A few school mementoes are shown at right which were collected by Mom and my sister Janet. The early-day school is similar to the old neighborhood Block School.

AGING GRACEFULLY. Like many of the students who spent time within the school's walls, this old tree has matured and endured.

one classmate was sent home early and ended up missing graduation as his punishment.

After we returned home, the school administration and school board brought an end to these annual class trips. The next senior class still got to go because they had already been earning trip money, but the decision was made that future senior classes would be staying home.

Out of 99 students in our graduating class, only a dozen would go directly on to college after high school graduation.

Most of the 50 boys in the class would work in one of the General Motors plants. At that time, General Motors auto plant work looked like a profitable career for life. But that wasn't to be as the industry changed drastically.

After I graduated from Michigan State University four years later and went back to my five-year high school class reunion the following year, I realized classmates working on the assembly line in the auto plants were earning more than twice as much money as I was making at the time.

Going back to the high school reunion really made me wonder about the value of a college education.

But I soon got over that feeling and realized the importance of doing what you really wanted to do and finding ways to grow in your career.

FRIDAY NIGHT ACTION. The football field and the basketball court were where almost everyone in town gathered on Friday nights through the 1950s.

82 Reasons Why Farming Is The Best Way Of Life

TIMES HAVE certainly changed since the Lessiter family started farming back in 1853.

Years ago, city folks snickered at those in the farm community and their simple way of life.

How could they survive without the opera, $100-plate restaurants, the bright lights and fast-paced action offered by the big city?

Not so anymore. Today, there's air pollution, overcrowdedness, sky-rocketing costs, crime, business people with ulcers the size of Mount St. Helens and nowhere for the kids to play and grow up safely.

As we've always said, there's no more rewarding, fruitful and hardworking calling in life than farming. Despite its problems and challenges, the serenity, peace and pride taken in farming is something people all over the globe are appreciating more and more every day.

As I was telling someone the other day, farming—as a way of

SUNDAY WALK IN THE FIELDS. Taking time off so three generations of the family can enjoy nature is among the many benefits of growing up on a farm.

life—represents all that's right about America. Most people don't become farmers—it's usually a trait born in everyone who's lucky enough to make a living in agriculture.

It's a part of a man's soul, gets in your blood and, in today's crazy world, is a slice of heaven that can be found in every rural town across this continent.

Here are 82 reasons why farming is better than any other way of life—hands down. I'm sure you'll think of even more.

1 No crowded subway, bus or slow-moving battle through bumper-to-bumper traffic. You grab your thermos, pet the dog and "foot it" across the gravel drive out to the barn.

2 Your office is one beautiful green acre after another.

3 Calling your most valued employees by the names of Bessie, Licorice and Big Red.

4 Your alarm clock consists of the amber-colored rooster and the tempting aroma of fresh-brewed coffee and steamy pancakes.

5 Raising your own Thanksgiving turkey.

6 Sharing in the key business decision-making process with your partner—your wife—with the lights out in bed.

7 In 100-degree weather, making an executive decision everyone can live with: grabbing an innertube, a cooler and heading to the pond out back.

8 Taking your steer to market or harvesting your crop and remembering the moment months earlier when you pulled the calf and planted the seed—creating new life from nothing.

9 Having your own private gas pump.

10 A view of the oak-lined drive and the waves of wind flowing through the tall grass.

11 Coming across something every day that reminds you of when you were 8 years old.

12 Never doing the same job two days in a row.

13 Your flannel shirt, baseball cap and overalls make up your outfit. No necktie to choke you.

14 You don't push papers, count numbers or stomp on the dreams of others—you feed the world.

15 The smell of the soil after a gentle autumn rain.

16 Vacationing for 10 days with your family nearly every winter.

17 No fax machines, gossipy secretaries or slick salesmen in three-piece suits to hound you.

18 Watching your wife labor diligently in her prized garden.

19 Escaping on hay rides with the church group in the brisk autumn air.

20 Having an unlocked front door.

21 Lots of time for solitary reflection—in the tractor seat, while fixing fenceposts, grinding feed, etc.

22 A refreshing shower after a tough but satisfying day of hard work and sweat.

23 Working a full day with the family dog by your side.

24 Getting lots of mud on your pants.

25 Sharing the tractor with

HORSING AROUND. Janet Roberts tosses fresh hay to the farm's band of horses.

your next-door neighbor.

26 Lying on your back in a hammock out in the yard watching three sets of fireworks from neighboring towns light up the sky on the 4th of July.

27 Early and late evening fishing.

28 Instead of pestering phone calls, you hear hearty moos and other animal greetings.

29 You work hard—but you've never had a day that wasn't rewarding or had to go to sleep without having pride in your heart.

30 Ties to the rural community—lifelong friends, sense of belonging in a certain area.

31 Watching your sons' and daughters' eyes light up the first time they see calves born.

32 Listening to the Saturday—and Sunday—afternoon football games as you harvest.

33 Following the traditional generation after generation footsteps which your family has for farming.

34 Wearing the many farming hats of business manager, nutritionist, agronomist, engineer, mechanic, veterinarian, construction worker, psychologist, salesman, etc.

35 Singing your favorite corny song—as loud as you wish— while you work.

36 Climbing into your grandfather's old sleigh for a ride on a frosty December night.

37 Making your own apple cider each fall.

38 Pitching horseshoes after lunch with the hired hands before returning to work.

39 Cutting your own baseball diamond with the lawnmower for the annual family game in July.

40 Knowing your kids are dreaming about owning their own spread—just like Mom and Dad.

41 The excitement in your blood every winter while preparing for spring.

42 Watching Grandpa tell the kids about the good 'ol days.

43 Teaching your dog to retrieve in the swamp.

44 Working closely with your family.

45 Enjoying three days at the State Fair—as a veritable business expense.

46 Stopping by your

UNCLE NEIL. Having never lived on a farm, Neil Roberts took to it in a big way after marrying Janet. His Simmental cattle and beagle pups won many honors.

friends' farm at any time of the day and always being asked to stay for dinner.

47 Knowing one day you'll pass this on to young Johnny and he will do the same for his children—for many generations to come.

48 Hanging the giant red ribbon around the whitewashed barn at Christmas.

49 Watching your grandchildren sing and play as you work.

50 Eating your wife's scrambled eggs, which you know were brought in from the henhouse less than a half-hour earlier.

51 Taking farm business risks with your wife.

52 Seeing the ear-to-ear smile on your son and daughter's face the first time they win a prize at the county fair.

53 Trading jokes at the feed mill or fertilizer dealership, sharing one-of-a-kind camaraderie and farm humor.

54 Fully enjoying the country's peace and quiet.

55 Building the kind of house you've always dreamed of—without city restrictions to make the decision for you.

56 Working on 4-H projects and events.

57 Performing the day to day loves, like pulling a calf.

58 Hiring your kids—and all their friends—for special projects and getting to know both parties a little better than before.

59 Watching snow gently kiss your land from your bedroom window.

60 Hunting with friends on your own land.

61 Knowing everyone at the diner, barber shop and grocery.

62 Making decisions on your farm's future yourself, without board room bureaucracy or office politics to stifle your innovation and creativity.

63 Owning your own place and business in the country.

64 Working the specific hours you want.

65 Giving away one of your rabbits to a 3-year-old city visitor every spring.

66 Picking a bouquet of wildflowers for your wife before coming in after a long day's work.

67 Hiding eggs the way the Easter Bunny intended for it to be done—outside.

68 Watching visitors stop by the side of the road and take pictures of your place. After you invite them in and give them the grand tour, you know they're wishing it was all theirs.

69 Taking your time on a Sunday afternoon.

70 Having freedom for some of life's unnecessary acts—like coaching Billy's grade school basketball team, leading the Scouts and 4-H, volunteering at the schools, organizing parties for neighbors and helping out the next-door neighbor with harvesting when he's in a real crunch.

71 Getting all "duded up" on a Sunday morning, going to church and then to the house of a friend you've known since kindergarten for a barbecue.

72 Driving down the farm lane with the wife and kids in October looking for deer.

73 Kneeling to clutch the moist, fruitful soil—the same thing your father and grandfather did many years before.

74 Growing pumpkins for the jack-o'-lanterns.

75 Saddling up a horse after dinner and riding down the lane as the sunset comes.

76 Having a rural route number on your mailbox.

77 Fishing without leaving home.

78 Getting the town kids in shape for football by having them haul hay.

79 Going into the city and enjoying it—knowing you'll soon be returning home.

80 When the last bushel is off to market, the most important benefit of being a farmer—time to enjoy life and all you've worked for with your family.

81 Getting to spend valuable time with Dad picking rocks out of fields.

82 Always having fresh milk—right from the cow—to drink at every meal.

45 Things I Remember Most About Growing Up On Lohill Farm

SPENDING 20 YEARS growing up on the family's farm in the 1940s and 1950s, I came away with enough favorable memories to last a lifetime.

While there's certainly nothing earthshaking or magical about any of the items on this list, they are just a few of the things that readily come to mind when I look back at the memorable times spent during my early years on our farm. These are things I would never trade for anything else.

THE FARM OFFICE. For many years, Dad did all of the farm's paperwork from this desk and the files kept in its drawers.

Here's hoping this list of memories from growing up at Lohill Farm also brings back some of your special times.

1 Dad leaning way out and spray painting the peak of the big barn with his legs wrapped around the third rung from the top of a 60-ft. extension ladder. I was sure glad to see him finish painting and get down!

2 Dad having dessert for breakfast. After milking for an hour or so in the morning, we'd come to the house for a big breakfast. Dad always topped off the early morning meal with a piece of pie or cake—the only person I know who always insisted on dessert for breakfast.

3 Being in seventh grade before I ever learned soup du jour wasn't french onion soup. As a farm boy, every time I ate out for years, I guess that's what they were serving that particular day.

4 Listening to Truth or Consequences on the radio at 8 p.m. on Saturday night while staying at Grandma and Grandpa Tarpening's house...and popping a great tasting batch of popcorn on the fast-heating gas stove.

5 After a big noon-time dinner on those dog days of summer, Dad would sometimes doze off while cultivating corn. Usually, it only meant a few young corn plants got clipped off.

But there was the day he fell asleep while cultivating near the lake and woke up to find the front wheels of the tractor only a couple of feet from the edge of the lake—which drops off to a 60-ft. depth only several feet offshore. He never fell asleep while cultivating that field again!

6 Having only one air conditioning unit on the farm—in the egg cooling room.

7 Squeezing grubs out of the backs of cows and squealing with laughter as they popped through the skin.

8 Flying with Dad when I was 7 years old from Detroit to Indianapolis (with stops in Toledo and Fort Wayne and throwing up all the way) to pick up a school bus.

Mom was expecting a baby at the time and Grandma Tarpening didn't think it was a very good time to be gone. When we got home that night, Dad spent 25 minutes maneuvering the school bus into the potato cellar so it would be protected overnight from the July elements.

9 Smelling 200-gallons of fuel oil spilled on the basement floor after Dad forgot to put the plug back in after draining the tank before ordering another load of winter fuel.

10 Looking at the postcards Grandma Norah had collected from winter trips to California and Florida.

11 Dad walking across the 12-inch wide beams 40-foot in the air with nothing underneath but the barn floor down below in the big barn. It used to really scare me to watch him rig the ropes for the hay forks.

12 How a New Year's Eve ice storm forced 18 of my parents' friends to spend the night...and how the next morning three went to the barn to watch my Dad milk while three others walked a half-mile to the country store to buy cigarettes.

13 Getting a telephone call that Mom's car had been rear-ended by another driver near Randall Beach. Dad and I quit baling hay and rushed over to find she wasn't hurt, but her pas-

GRIGGS DRUGSTORE. The brick building on the left in downtown Lake Orion used to house one of the town's two drugstores, which were directly across the street from one another.

senger, Mrs. Waters who lived with my Grandmother, had hit her head on the dashboard (way before seatbelts were required) and had to go to the hospital.

The man who rear-ended Mom's car didn't have a driver's license. He tried to convince the sheriff that even though he was sitting next to the driver's side window, it was his girlfriend sitting close to him who was doing the driving. It didn't work.

14 Going early to the Saturday afternoon movies at the Lake Orion theatre to help the truck driver haul in the film cans for next week's shows.

15 Getting my legs so sunburned on a 3-day 4-H canoe trip that I couldn't get up off the davenport for the next 5 days.

16 Looking through old livestock breed association magazines and having Dad tell me he'd visited nearly all the advertised farms many years earlier when he was a member of the Michigan State College livestock judging team.

17 Seeing my sister get a concussion when she ran into a chain in a cattle alleyway and having it flip her upside down, hitting her head on the concrete. She ended up in the hospital.

18 Remembering when Mom only washed clothes on Mondays and only ironed on Tuesdays.

19 Having parents with the very unusual names of Milon (Dad's grandfather's name which he hates—so he goes by John) and Donalda (named after her father, Donald, who died 10 days before she was born).

20 Going up to Grandma Norah's house to play rummy with her.

21 Having spent at least 30 days in the dorm at Michigan State University during various state 4-H events and then living in the same dorm when I went to college.

22 Dropping full paper bags of water out the tenth story windows of the downtown Detroit hotel while aiming at people walking along the streets during the annual Detroit Junior Livestock Show.

23 Seeing the lake level drop 6 inches one hot dry summer when the neighboring farm irrigated potatoes practically every hour of every day.

24 Seeing my dad for nearly a decade become almost a professional pallbearer month after month as some of my grandparents' friends died.

25 Dressing up in a little sailor suit to look like my Uncle Sherm, who served on one of the Navy's destroyers in the Pacific Ocean during World War II.

26 Going to Canada to look for lambs that could whip everyone else's entries at the Detroit Junior Livestock Show.

27 Leaving a gallon jug of cold water under a wheat shock to keep it shaded and cool, then coming back an hour later unable to remember which shock it was under.

28 Sharing a 16-person telephone party line and hearing the hired man's wife ask Mom about something that could only have come from listening to a telephone conversation.

29 Watching Dad take a 20-minute nap after dinner, then get up at 1 p.m. wide awake and

LONG GONE. As a kid, *Farm Quarterly* was among my favorite farm magazines, with its in-depth articles on farming.

ready to tackle another afternoon's farm work.

30 Never building a fire in the fireplace because the thermostat was directly across the room so the whole house would be cold. And years later, smiling when Mom had to explain to our kids when they asked why the bricks were painted green, why the piano was sitting in front of the fireplace and why they'd never seen a fire burning there.

31 Swinging a full basket of eggs or a pail of water through a 360-degree circle without breaking an egg or spilling a single drop of water.

32 Going to the funerals of Great Grandmother Wiser and later Grandpa Frank in the living room of the big house—not at the funeral home in town. And seeing the room totally jammed with flowers and the yard full of parked cars.

33 Spending 50 cents on a Saturday afternoon at the movie house in Lake Orion—10 cents for a ticket, 10 cents for a box of popcorn, 10 cents for a Pepsi, 10 cents for another box of popcorn and 10 cents for another Pepsi.

34 Being amazed when the car dealer dropped a $20 bill into the collection plate every Sunday morning. And later finding he spent several years in jail for defrauding the Internal Revenue Service.

35 Seeing Dad fall through the slatted floor in the poultry house and watching a full basket of eggs break.

36 When snow drifted over the 25-ft. tall peak on the neighbor's sheep barn next to the road, a stranger stayed with us for 3 days. The road crew ended up hooking two road graders together to clear a path after the big 1947 blizzard.

37 Growing a beard when I came home from college for the summer and having Mom tell me she didn't like it. About a week later, Dad and I came in for lunch after a hard morning of lugging hay bales and there was nothing ready to eat. When asked why, Mom said, "There won't be any food until Frank shaves."

Dad and I were so hungry that I went upstairs and shaved. In 15 minutes, lunch was ready!

38 Remembering how mad all the 4-H Club leaders were when chicken was served at the Michigan Junior Livestock Show banquet in Detroit to kids who had worked so hard to produce beef, lamb and pork. Although the banquet was always planned by city folks in Detroit, that never happened again.

39 Getting to wear all the sailor caps my Uncle Sherm brought home from the Navy after World War II.

40 Feeding and watering the chickens and gathering eggs every day after school when I was only 8-years-old.

41 Visiting Grandma Norah when the Watkins or Cook Coffee man showed up and watching with keen interest as they opened their huge display cases and showed off all of their products.

42 Having a former english teacher for a mother who greatly encouraged my writing skills.

43 Shoveling coal from the bin into the stoker, removing the clinkers and adding water to the boiler at Grandma Norah's house every afternoon after school.

44 Getting a big toy steam shovel for Christmas and having Dad lug in several bushels of wheat and dumping it on the basement floor so I could play with it. And heading for the hospital 2 days later for a hernia operation when I was only 6 years old.

45 Riding the old rickety streetcars 8 miles from the downtown Detroit hotels to the Michigan State Fairgrounds twice a day during the annual 3-day Detroit Junior Livestock Show.

DAIRY COWS AND MILK. That's how the family farm made its income over a number of years.

Family Favorites

A GRANDMOTHER'S wedding ring, an antique chest, perhaps a gold pocket watch—these are only a few of the farm family keepsakes which are handed down from generation to generation.

In the Lessiter family, favorite recipes have always been among those highly treasured items that have been passed down. Here are some of our favorites from a baker's dozen of family members spanning four generations.

Apple Crunch

6 large juicy apples
1 1/2 c. sugar
3/4 c. butter or shortening (not oleo)
1 c. flour
2 tsp. lemon juice (over apples)
1 1/2 tsp. cinnamon
1/2 tsp. nutmeg

Peel and quarter apples. Place in greased 9-inch cake pan. Pour lemon juice over. Mix other ingredients and pat over top of apples. Bake at 375 degrees for 1 hour.

—*Janet Roberts*

Carrot Cake

Sift:
2 c. flour
1 tsp. baking powder
2 tsp. soda
2 tsp. cinnamon

Mix:
2 c. white sugar
1 1/2 c. oil
3 c. grated carrots
Nutmeats

Add dry ingredients, then add 4 eggs one at a time. Bake at 350 degrees for 45 to 60 minutes.

Carrot Cake Frosting...
1 8 oz. cream cheese (softened)
1/2 stick butter
1 box powdered sugar
1 tsp. vanilla

—*John Lessiter*

Chocolate Chip Cookies

Sift:
    3 3/8 c. flour
    3/4 tsp. baking soda
    1 1/2 tsp. salt

Cream:
    3/4 c. brown sugar
    1 1/2 c. sugar
    1 1/2 c. shortening

Add:
    3 eggs
    3 tsp. vanilla
    Chocolate chips

Mix together. Bake at 350 degrees for 10 minutes.

—*Frank Lessiter*

Christmas Sugar Cookies

1 c. soft shortening
1 1/2 c. sugar
1/2 t. salt
2 eggs unbeaten
1 tsp. vanilla
1/2 tsp. almond extract

Mix until light and fluffy.

Dissolve:
    1/2 tsp. baking soda
    3 Tbsp. sour or sweet cream

Add to other mixture and blend in 3 cups of sifted flour.

Chill dough, roll and cut. Bake at 375 degrees for 8 to 10 minutes.

—*Susie Grabow*

Hot Fudge Sauce

1 c. sugar
1/4 c. flour
3 Tbsp. cocoa or 2 squares chocolate
1 1/3 c. milk (warm)

Mix and cook until thick. Add 1/4 tsp. salt, 1/2 tsp. vanilla and 1 Tbsp. butter.

—*Neil Roberts*

Lemon Cake

1 pkg. yellow cake mix
1 pkg. lemon pie filling
4 eggs
3/4 c. oil
3/4 c. water

Combine and beat for 5 minutes. Bake in 9 x 13-inch pan for 45 minutes at 350 degrees. Remove from oven and pierce all over with ice pick or large fork. Then pour syrup over hot cake.

Blend for Syrup:
    2 c. powdered sugar
    1/4 c. butter
    1/2 c. lemon juice

—*Trista Linman*

Mandarin Molded Salad

Dissolve:
    1 pkg. orange Jello
    1 pkg. lemon Jello
    1 c. boiling water

Add:
    1 small can frozen orange juice
    Juice from 11 oz. can mandarin orange

Let mixture set until syrupy. Put mandarin orange in mold.

Pour 1/3 liquid over them and let sit until firm.

Whip 2 packages dream whip and add a pinch of salt.

Fold syrupy Jello in until foamy. Pour over orange mixture. Let set until firm.

—*Katie Roberts*

Pistachio Fluff

1 small box instant pistachio pudding
1 can (#2) crushed pineapple (don't drain)
1 8 oz. tub whipped topping
1 c. miniature marshmallows

Empty pudding mix into bowl. Add pineapple and juice and mix. Fold in whipped topping and marshmallows. Mix together and let stand for 5 minutes.

—*Kelly Lessiter*

SOME GREAT FARM COOKING. Donalda Lessiter, left, prepares another excellent, yet quick, mid-summer family supper that includes hotdogs, baked beans, homemade pickles and freshly picked tomatoes from the farm garden for family and visitors. Enjoying the meal are Dick Reese, Janet, John and Frank Lessiter.

Pineapple Upside Down Cake

1 c. brown sugar
1/4 lb. butter
1 can sliced pineapple
1 c. sugar
5 tbsp. pineapple juice
3 egg yolks
3 egg whites, beaten
1 c. flour
1 Tbsp. baking powder

Melt butter in skillet and spread on brown sugar. Then lay sliced pineapple on the sugar. Beat egg yolks, add sugar and pineapple juice. Sift flour and baking powder and add. Fold in beaten egg whites. Pour over the first mixture and bake in oven at 350 degrees. When done, place cake plate on top of pan and reverse. Serve with whipped cream.

—*Florence Foster Lessiter*

Meat Loaf

Mix:
1 1/2 lb. ground beef
1 1/2 lb. lean ground pork
1 c. cracker crumbs
2 eggs beaten
2 c. milk
1 tsp. salt
1/4 tsp. pepper
1 Tbsp. chopped onion
1/8 tsp. allspice

Bake in loaf pan for 1 hour and 20 minutes at 350 degrees.

—*Mike Lessiter*

Pork Chop Special

Brown 6 pork chops. Place in 9 x 13-inch dish. Season. Place a spoonful of cooked rice on top of each chop. Add a slice of onion. Cover with tomatoes or tomato soup. Bake at 350 degrees for 1 1/2 hours. Add additional tomatoes if needed.
— *Pam Lessiter*

No-Bake Cookies

Bring to rolling boil, boil for 1 minute:
    2 c. sugar
    1/2 c. milk
    1/2 c. oleo
    1/2 c. cocoa (optional)

Add:
    1 c. chunky peanut butter
    2 tsp. vanilla

Beat until smooth.

Add:
    3 c. oats
    1 c. peanuts

Cool and place on waxed paper.
— *Andy Roberts*

7-Layer Cookies

    1 stick oleo

Melt oleo in 9 x 13-inch pan.

Sprinkle over it:
    1 c. graham cracker crumbs
    1 c. coconut
    1 small package chocolate chips
    1 small package butterscotch chips

Drizzle over this one 15-ounce can of sweetened condensed milk. Sprinkle 1 1/2 cups chopped nuts over this. Bake at 350 degrees for 30 minutes. Cut and cool.
— *Debbie Hansen*

AFTER CHRISTMAS DINNER. Games were always popular after a big family meal. In this special three-generation Christmas Day gathering, Sherman Tarpening plays a brand new game with grandchildren Steve Tarpening and Janet Lessiter. Beaming with approval in the background is Steve's mother, Helen Tarpening.

Orion Township— 1819 to 1880

HERE ARE A few of the early-day occurrences in Orion township during its first 60 years of existence:

1819—First land purchase in area by Judah Church and John Wetmore.

1825—Moses Munson was first settler in area. Built first saw mill and log cabin for family.

1825—Church services held by pioneer missionaries Norton, Warren and Earle.

1828—First cemetery was Bigler Cemetery.

1828—Early-day settlers included Munson, Carpenter, Decker, Bigler, Pinckney, Simmons, McVean, Bagg and McElvery families.

1829—Lake Canandaigua formed by dam across Paint Creek built by Needham Hemingway, Jesse Decker and Philip Bigler. They built a saw mill below the dam and a log house for the sawyer which later became a tavern owned by Thomas Abernathy.

1832—Rufus Streator builds blacksmith shop.

1832—First post office in township.

1832—Saw mill near Paint Creek was burned by Indians when tavern proprietor would not sell them whiskey.

1833—Society of Congregationalists formed at home of Needham Hemingway.

1834—John Haukison builds two-story store and storehouse.

1834—First log school built at Clark's Corner.

1835—Area was part of Oakland Township in 1820, became part of Pontiac Township in 1828 and was organized as Orion Township in 1835. Name was suggested by Jesse Decker who was impressed by beauty of the Orion name found in an old school book and urged its choice at first town meeting held in his home on April 5, 1835.

1835—The area was known as Canandaigua, New Canandaigua and Dogway by older residents.

1836—City of Canandaigua was platted by traveling auctioneer James Stilson on June 1, 1836. Lake was to be known as Lake Canandaigua. Lake lots were sold to easterners who departed in disgust after seeing land covered by marsh and brush. The remaining lake lots were sold to Detroiters for bargain price of only six cents per lot.

1836—With a sharp increase in the area's population from 1826 to 1836, food was scarce.

Pontiac's Dr. William saved the day by purchasing flour and corn from Ohio and delivering it to the rich and poor in Orion Township. Each person received four pounds of flour and a peck of corn.

1836—First grist mill, later known as Wilder's Mill, erected by Powell Carpenter.

1836—Township divided into four districts.

1837—Stores and offices are moved from Deckerville to current Village of Lake Orion.

1837—Needham Hemingway builds large flour mill in Orion village, powered by a raised and strengthened dam.

1837—Dr. Smead becomes area's first physician.

1838—Evergreen Cemetery was platted.

1838—Needham Hemingway plats 30 blocks in village. John Perry plats 16 blocks.

1844—First school house erected in village.

1844—First steam mill operated by Robert Merrick in Mahopac, an area in the southwest portion of the township.

1844—Formation of Presbyterian Society.

1849—Organization of I.O. and O.F. Lodges.

1852—Organization of Sons of Temperance.

1854—Official post office name changed from Canandaigua to Orion. Semi-weekly mail delivery offered.

1854—Society of Congregationalists church built, later known as Baptist Church.

1858—Jesse Decker plats 13 blocks in village.

1859—Orion officially incorporated as village.

1862—Nearly all of the village was destroyed by fire.

1863—Village charter is repealed.

1865—First purebred stock in township included five Shorthorn cattle purchased from New York by John Lessiter.

1869—The village was again rechartered.

1870—First Sunday School organized by Vincent Brown who served as its superintendent for many years.

1872—Michigan Central Railroad lays track through village and township.

1872—Methodist church was built.

1874—Organization of Good Templars.

1874—E.R. Emmons develops park on north side of lake and purchases the "Little Dick" steamer to carry passengers to islands on lake.

1874—A dozen prominent citizens purchase Park Island and build bridge to mainland. They construct dance hall and reception hall, purchase a steamer and add a wharf to bring people to and from the island.

1880—Park Hotel is built by Stephen Seeley.

—Presented by Norah Lessiter on September 10, 1925, at 100th anniversary party celebrating the formation of Orion Township.

Photo, Illustration Credits...

All of the photos, maps, land plats and illustrations appearing in this book came from the collections of Lessiter family members, except for the following:

Aerial Graphics—9 (top).
Aerial Photographers, Inc.—188.
Aker—325 (top).
Aberdeen Angus Journal—255, 256, 2157 (bottom), 258, 259, 260 (top), 161, 262.
Associated Press/World Wide Photos—254.
Breeders Gazette—206, 207.
Ford Archives, Henry Ford Museum—149 (top), 150, 328, 329, 330.
Ford Motor Company—327.
Iowa State University—250 (bottom).
Keehoos Studios—18, 19, 22, 29, 111, 128, 241, 253 263, 265, 275, 301, 305 (top), 321, 340 (top), 341, 349, 358 (top), 360, 362. 367.
Kellerhels—309.
Kot, Gregg—6 (bottom), 26, 27, 28.
Lake Orion Review—111.
Lakefield Farms—310 (top), 323,
National Automotive History Collection, Detroit Public Library—169, 170.
New York Central Railroad Collection of Allen County, Ohio, Historical Society—215, 216, 217, 218, 219, 220.
O.F. Foster family—310 (top and bottom left),
Oakland County, Michigan, Road Commission—156.
Pontiac Daily Press—160. 128.
Soil Survey Of Oakland County, Michigan—9 (bottom).
State Archives Of Michigan—66, 202, 203.
State Historical Society Of Wisconsin—67.
Soil Conservation Service—158.
Union Pacific Railroad—221.
United States Army—270 (bottom).
United States Navy—306 (bottom center).
Wetmore and Agri Graphics, Inc.—319, 320, 321, 322, 324, 325, 326.
Zekan-Robbins—6 (top).

200 Years Of Family History

AS BEST CAN BE pieced together from passed down memories, historical records and old family albums, here is the history of the Lessiter family.

The family's records trace the Lessiter family back as far as Wiltshire, England, in the early 1800s.

Since that time, there have been seven generations and six of these generations have actually lived on the Lessiter family farm west of Lake Orion, Michigan. The farm got its start in 1853 when it was settled by a member of the clan who made the long boat trip from England.

First Generation

Born in the early 1800s, William Lessiter married Elizabeth Kington on March 25, 1824. They had four sons, John, Henry, William and James, and a daughter, Elizabeth, who died at age 12.

After Willliam's wife, Elizabeth, died on March 11, 1833, he married Elizabeth Sheppard in October, 1838.

In 1843, William emigrated to the United States and began farming near the little town of Grettan located near Grand Rapids, Michigan.

FAMILY HISTORIAN. John Lessiter was the main source of information on the family for the past 200 years.

Second Generation

John Lessiter was born on July 19, 1827. Following the death of his mother when he was 5 years old, his father kept him in boarding school in England until he was age 12.

Since British tradition called for only one son to inherit all of

"British tradition called for the eldest son to take over the farm..."

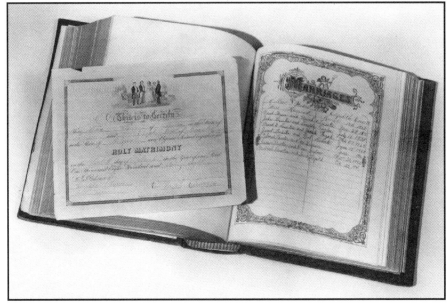

FAMILY BIBLE. Much of the history of the family is written on pages in the Bible which has been handed down from generation to generation.

the farmland, John Lessiter, at the age of 20, and his brother, James, made the six-week voyage by boat to the United States in 1847.

James ended up on a farm near the outskirts of Pontiac, Michigan, while John journeyed 10 miles north to the Lake Orion, Michigan, area, leaving behind a sweetheart in England who was a dressmaker by trade, with the promise that he would eventually send for her.

After working by the month for several Michigan landowners, John Lessiter bought his employer's livestock and rented a farm for several years. He purchased 120 acres in 1852 at the present farm's location.

On New Year's Day in 1853, he married Nancy Bearsley of nearby Clarkston, Michigan. They had six children, and their sons, Frank and Floyd, farmed with their Dad. The four girls included Libby, Edna, Ida Mae and Maggie, who died when she was only 12 years old.

Soon after starting the farming operation in 1853, the family became involved in raising and breeding Shorthorn cattle. They raised grade cattle for seven years. Then in 1865, they bought

SUPER HOME. When Frank and Floyd Lessiter split up the farm at the time of their father's death in 1901, this became the Floyd Lessiter family home. Soon after, Frank built a new home on the original farmstead.

FIVE GENERATIONS. Norah and Frank Lessiter would be 15-month-old Casey Grabow's great-great grandparents.

THE LESSITER CLAN. Clockwise from upper right: John Lessiter as a 7-month old baby with mother Norah, grandmother Elizabeth Wiser and great grandmother Lucy Everett, who used to sit behind the kitchen stove and smoke a clay pipe; Frank Lessiter as a one-year-old youngster; John Lessiter at the age of 16 months; two-year-old John Lessiter with grandfather Milon Wiser; and 10-year-old John with his parents, Norah and Frank Lessiter at upper left.

five head of registered Durham cattle from New York to start a purebred Shorthorn program.

The family also had a flock of registered Shropshire sheep.

"Industrious and thrifty, John Lessiter accumulated considerable money," according to a write-up in the 1912 *History Of Oakland County*. "When ready to make a permanent settlement, he bought 120 acres of land in Orion Township in 1852 where he subsequently resided until his death which occurred on October 23, 1901."

Third Generation

Born on February 6, 1862, Frank was the link among the six children to the present Lessiter family.

At the age of 33 on March 28, 1895, Frank married Norah Wiser who lived on a farm near Seymour Lake, Michigan. Norah was the only child of Elizabeth and Milon Wiser.

She grew up on a farm at Seymour Lake which had been purchased in 1831 and also later became a Centennial Farm for the family—one of only a few Michigan families who ever had two Centennial Farms in the

FOUR GENERATIONS. Elizabeth Wiser, John Lessiter, Norah Lessiter and nine-month-old Frank Lessiter.

FOUR GENERATIONS. Debbie, Molly, Ryan and Alex Hansen with grandfather and great grandfather John Lessiter.

family.

Milon Wiser had come from New York when he was 15 years old, and farmed 80 acres at Seymour Lake, located seven miles north of the Lessiter farm.

When their dad died in 1901,

> *"When they split up the farm, Grandpa Frank moved back to the original farmstead..."*

Frank, who was 39, and Floyd, who was 38, took over the 436-acre farming operation. Between 1900 and April 6, 1909, when their mother, Nancy, died, they bought out the remaining interest in the farm from their sisters.

Soon after, the brothers split the acreage and Frank moved a half mile north to the original Centennial farmstead.

Floyd continued to live in the old house which still stands today at the corner of Baldwin and Clarkston Roads. Later, he sold this land in the mid 1920s to Oscar Webber of Detroit and it became part of the huge Lakefield Farms operation.

After the sale, the family moved a mile west on Clarkston Road and developed a new farming operation. The price offered by Webber for the land surrounding Dennis Lake was just too good to pass up back in the 1920s.

While the two brothers frequently worked closely together, each had his own farming operation. Frank later bought a nearby tract of 120 acres. The big 40- by 96-foot barn was built in 1905 and is still going strong.

The 1912 *History Of Oakland County, Michigan,* refers to the farm as the "The Maples—with 308 acres with substantial buildings." However, members of the family alive at that time can never recall the farm being called by that name.

The book went on to state that, "Possessing good judgment and sound sense, he (Frank) succeeded well in his agricultural work. He added to his land and possessions until he had a farm of 436 acres—all paid for and in a fair state of tillage.

"In addition to carrying on general farming with most satisfactory results, Frank Lessiter was one of the leading stock raisers of the township, having a fine herd of registered Shorthorn cattle. The head of the herd was a bull named Oakland Prince, a registered yearling bull.

"The farm had large maple and pine trees bordering the road and a large lake at the rear of the house. There is a large modern steam-heated house. Also there is a tenant house—the first home built by John Lessiter which has since been remodeled.

"Frank Lessiter was an adherent of the Democratic party and served as township treasurer. He was a member of the ancient free and accepted order of Masons, the order of the Eastern Star, the Benevolent and Protective Order of Elks, the Knights of the Maccabees and of the ancient order of Mayners. Religiously, he was a Methodist."

Frank was educated at district schools and Pontiac Business College. But much of his knowledge of agriculture actually came from working alongside his father during his youth; and his father proved to be a wise instructor.

Fourth Generation

The next member of the Lessiter family, John, was born on June 25, 1908. Another child, Caroline, born to Norah and Frank Lessiter died at birth.

John rode his pony to school at the Block School, about a mile from the farm. During high school, he lived with his grandmother in Oxford during the week and came back to the farm on weekends. By the time he was a senior, he had a Chevrolet roadster and drove back and forth to school each day.

John recalls spending many exciting days with the families of his five aunts and uncles. Several come to mind as favorites.

He spent considerable time playing and working with his cousins Bruce and Marion, children of Floyd and Lillian Lessiter, who lived only about a mile away.

TWO FARM GENERATIONS. Donalda and John Lessiter along with John's parents, Norah and Frank Lessiter.